Lecture Notes in Artificial In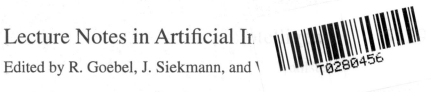

Edited by R. Goebel, J. Siekmann, and \

Subseries of Lecture Notes in Computer Science

François Fages   Francesca Rossi
Sylvain Soliman (Eds.)

Recent Advances in Constraints

12th Annual ERCIM International Workshop
on Constraint Solving and Constraint Logic Programming,
CSCLP 2007
Rocquencourt, France, June 7-8, 2007
Revised Selected Papers

 Springer

Series Editors

Randy Goebel, University of Alberta, Edmonton, Canada
Jörg Siekmann, University of Saarland, Saarbrücken, Germany
Wolfgang Wahlster, DFKI and University of Saarland, Saarbrücken, Germany

Volume Editors

François Fages
Sylvain Soliman
INRIA Rocquencourt
Projet Contraintes
Domaine de Voluceau, BP105, 78153 Le Chesnay CEDEX, France
E-mail: {Francois.Fages, Sylvain.Soliman}@inria.fr

Francesca Rossi
Università di Padova
Dipartimento di Matematica Pura ed Applicata
Via Trieste 63, 35121 Padova, Italy
E-mail: frossi@math.unipd.it

Library of Congress Control Number: 2008940263

CR Subject Classification (1998): I.2.3, F.3.1-2, F.4.1, D.3.3, F.2.2, G.1.6, I.2.8

LNCS Sublibrary: SL 7 – Artificial Intelligence

ISSN   0302-9743
ISBN-10   3-540-89811-5 Springer Berlin Heidelberg New York
ISBN-13   978-3-540-89811-5 Springer Berlin Heidelberg New York

Springer is a part of Springer Science+Business Media

springer.com

© Springer-Verlag Berlin Heidelberg 2008
Printed in Germany

Typesetting: Camera-ready by author, data conversion by Scientific Publishing Services, Chennai, India
Printed on acid-free paper   SPIN: 12568338   06/3180   5 4 3 2 1 0

Preface

Constraint Programming supports a great ambition for computer programming: the one of making of programming essentially a modeling task, with equations, constraints and logical formulas. This field emerged in the mid eighties borrowing concepts from Logic Programming, Operations Research, and Artificial Intelligence. Its foundation is the use of relations on mathematical variables to compute with partial information systems. The successes of Constraint Programming for solving combinatorial optimization problems in industry or commerce are related to the advances made in the field on new constraint propagation techniques and on declarative languages which allow control on the mixing of heterogeneous resolution techniques such as numerical, symbolic, deductive and heuristic.

This volume contains the papers selected for the post-proceedings of the 12th International Workshop on Constraint Solving and Constraint Logic Programming (CSCLP'07) held on June 7-8th 2008 in Rocquencourt, France. This workshop, open to all, was organized as the twelfth meeting of the working group on Constraints of the European Research Consortium for Informatics and Mathematics (ERCIM), continuing a series of workshops organized since the creation of the working group in 1997. A selection of papers of these annual workshops are published since 2002 in a series of books which illustrate the evolution of the field, under the title "Recent Advances in Constraints" in the Lecture Notes in Artificial Intelligence.

This year, there were 16 submissions, most of them being extended and revised versions of papers presented at the workshop, plus some new papers. Each submission was reviewed by three reviewers. The programme committee decided to accept 10 papers for publication in this book.

We would like to take the opportunity to thank all authors who submitted a paper, as well as the reviewers for their useful work. CSCLP'07 has been made possible thanks to the support of the European Research Consortium for Informatics and Mathematics (ERCIM), the Institut National de la Recherche en Informatique et Automatique (INRIA) and the Association for Constraint Programming (ACP).

May 2008   François Fages, Francesca Rossi, and Sylvain Soliman

Organization

CSCLP 2007 was organized by the ERCIM Working Group on Constraints.

Organizing and Program Committee

François Fages   INRIA Rocquencourt, France
Francesca Rossi University of Padova, Italy
Sylvain Soliman INRIA Rocquencourt, France

Additional Reviewers

Erika Abraham
Krzysztof Apt
Pedro Barahona
Roman Barták
Thierry Benoist
Hariolf Betz
Stefano Bistarelli
Ismel Brito
Ondrej Cepek
Marco Correia
Yves Deville
Gregoire Dooms
Thom Frühwirth
Martin Fränzle
Khaled Ghedira
Daniel Goossens
Martin Hejna

Christian Herde
Boutheina Jlifi
Ulrich Junker
Chavalit Likitvivatanavong
Pedro Meseguer
Barry O'Sullivan
Gregory Provan
Igor Razgon
Francesco Santini
Pierre Schaus
Pavel Surynek
Tino Teige
Kristen Brent Venable
Ruben Viegas
Roland Yap
Stéphane Zampelli

Sponsoring Institutions

European Research Consortium for Informatics and Mathematics (ERCIM),
the Institut National de la Recherche en Informatique et Automatique (INRIA),
and the Association for Constraint Programming (ACP)

Table of Contents

A Comparison of the Notions of Optimality in Soft Constraints and Graphical Games

Krzysztof R. Apt[1,2], Francesca Rossi[3], and K. Brent Venable[3]

[1] CWI Amsterdam, Amsterdam, The Netherlands
[2] University of Amsterdam, Amsterdam, The Netherlands
[3] University of Padova, Padova, Italy
apt@cwi.nl, {frossi,kvenable}@math.unipd.it

Abstract. The notion of optimality naturally arises in many areas of applied mathematics and computer science concerned with decision making. Here we consider this notion in the context of two formalisms used for different purposes and in different research areas: graphical games and soft constraints. We relate the notion of optimality used in the area of soft constraint satisfaction problems (SCSPs) to that used in graphical games, showing that for a large class of SCSPs that includes weighted constraints every optimal solution corresponds to a Nash equilibrium that is also a Pareto efficient joint strategy.

We also study alternative mappings including one that maps graphical games to SCSPs, for which Pareto efficient joint strategies and optimal solutions coincide.

1 Introduction

The concept of optimality is prevalent in many areas of applied mathematics and computer science. It is of relevance whenever we need to choose among several alternatives that are not equally preferable. For example, in constraint optimization, each solution of a constraint problem has a quality level associated with it and the aim is to choose an optimal solution, that is, a solution with an optimal quality level.

The aim of this paper is to clarify the relation between the notions of optimality used in game theory, commonly used to model multi-agent systems, and soft constraints. This allows us to gain new insights into these notions which hopefully will lead to further cross-fertilization among these two different approaches to model optimality.

Game theory, notably the theory of *strategic games*, captures the idea of an interaction between agents (players). Each player chooses one among a set of strategies, and it has a payoff function on the game's joint strategies that allows the player to take action (simultaneously with the other players) with the aim of maximizing its payoff. A commonly used concept of optimality in strategic games is that of a Nash equilibrium. Intuitively, it is a joint strategy that is optimal for each player under the assumption that only he may reconsider his action. Another concept of optimality concerns Pareto efficient joint strategies,

F. Fages, F. Rossi, and S. Soliman (Eds.): CSCLP 2007, LNAI 5129, pp. 1–16, 2008.

which are those in which no player can improve its payoff without decreasing the payoff of some other player. Sometimes it is useful to consider constrained Nash equilibria, that is, Nash equilibria that satisfy some additional requirements [6]. For example, Pareto efficient Nash equilibria are Nash equilibria which are also Pareto efficient among the Nash equilibria.

Soft constraints, see e.g. [2], are a quantitative formalism which allow us to express constraints and preferences. While constraints state what is acceptable for a certain subset of the objects of the problem, preferences (also called *soft constraints*) allow for several levels of acceptance. An example are fuzzy constraints, see [4] and [11], where acceptance levels are between 0 and 1, and where the quality of a solution is the minimal level over all the constraints. An optimal solution is the one with the highest quality. The research in this area focuses mainly on algorithms for finding optimal solutions and on the relationship between modelling formalisms (see [9]).

We consider the notions of optimality in soft constraints and in strategic games. Although apparently the only connection between these two formalisms is that they both model preferences, we show that there is in fact a strong relationship. This is surprising and interesting on its own. Moreover, it might be exploited for a cross-fertilization among these frameworks.

In considering the relationship between strategic games and soft constraints, the appropriate notion of a strategic game is here that of a *graphical game*, see [7]. This is due to the fact that soft constraints usually involve only a small subset of the problem variables. This is in analogy with the fact that in a graphical game a player's payoff function depends only on a (usually small) number of other players.

We consider a 'local' mapping that associates with each soft constraint satisfaction problem (in short, a soft CSP, or an SCSP) a graphical game. For strictly monotonic SCSPs (which include, for example, weighted constraints), every optimal solution of the SCSP is mapped to a Nash equilibrium of the game. We also show that this local mapping, when applied to a consistent CSP (that is, a classical constraint satisfaction problem), maps the solutions of the CSP to the Nash equilibria of the corresponding graphical game. This relationship between the optimal solutions and Nash equilibria holds in general, and not just for a subclass, if we consider a 'global' mapping from the SCSPs to the graphical games, which is independent of the constraint structure.

We then consider the relationship between optimal solutions of the SCSPs and Pareto efficiency in graphical games. First we show that the above local mapping maps every optimal solution of a strictly monotonic SCSP to a Pareto efficient joint strategy. We then exhibit a mapping from the graphical games to the SCSPs for which the optimal solutions of the SCSP coincide with the Pareto efficient joint strategies of the game.

In [5] a mapping from graphical games to classical CSPs has been defined, and it has been shown that the Nash equilibria of the games coincide with the solutions of the CSPs. We can use this mapping, together with our mapping from the graphical games to the SCSPs, to identify the Pareto efficient Nash equilibria of the given graphical game. In fact, these equilibria correspond to the optimal solutions of the SCSP obtained by joining the soft and hard constraints generated by the two mappings.

The study of the relations among preference models coming from different fields such as AI and game theory has only recently gained attention. In [1] we have considered the correspondence between optimality in CP-nets of [3] and pure Nash equilibria in so-called parametrized strategic games, showing that there is a precise correspondence between these two concepts.

As mentioned above, a mapping from strategic, graphical and other types of games to classical CSPs has been considered in [5], leading to interesting results on the complexity of deciding whether a game has a pure Nash equilibria or other kinds of desirable joint strategies.

In [12] a mapping from the distributed constraint optimization problems to strategic graphical games is introduced, where the optimization criteria is to maximize the sum of utilities. By using this mapping, it is shown that the optimal solutions of the given problem are Nash equilibria of the generated game. This result is in line with our findings regarding strictly monotonic SCSPs, which include the class of problems considered in [12].

2 Preliminaries

In this section we recall the main notions regarding soft constraints and strategic games.

2.1 Soft Constraints

Soft constraints, see e.g. [2], allow to express constraints and preferences. While constraints state what is acceptable for a certain subset of the objects of the problem, preferences (also called *soft constraints*) allow for several levels of acceptance. A technical way to describe soft constraints is via the use of an algebraic structure called a c-semiring.

A *c-semiring* is a tuple $\langle A, +, \times, \mathbf{0}, \mathbf{1} \rangle$, where:

- A is a set, called the **carrier** of the semiring, and $\mathbf{0}, \mathbf{1} \in A$;
- $+$ is commutative, associative, idempotent, $\mathbf{0}$ is its unit element, and $\mathbf{1}$ is its absorbing element;
- $\times$ is associative, commutative, distributes over $+$, $\mathbf{1}$ is its unit element and $\mathbf{0}$ is its absorbing element.

Elements $\mathbf{0}$ and $\mathbf{1}$ represent, respectively, the highest and lowest preference. While the operator $\times$ is used to combine preferences, the operator $+$ induces a partial ordering on the carrier A defined by

$$a \leq b \text{ iff } a + b = b.$$

Given a c-semiring $S = \langle A, +, \times, \mathbf{0}, \mathbf{1} \rangle$, and a set of variables V, each variable x with a domain $D(x)$, a **soft constraint** is a pair $\langle \text{def}, \text{con} \rangle$, where con $\subseteq V$

and def : $\times_{y \in \text{con}} D(y) \to A$. So a constraint specifies a set of variables (the ones in con), and assigns to each tuple of values from $\times_{y \in \text{con}} D(y)$, the Cartesian product of the variable domains, an element of the semiring carrier A.

A **soft constraint satisfaction problem** (in short, a **soft CSP** or an SCSP) is a tuple $\langle C, V, D, S \rangle$ where V is a set of variables, with the corresponding set of domains D, C is a set of soft constraints over V and S is a c-semiring. Given an SCSP, a **solution** is an instantiation of all the variables. The **preference** of a solution s is the combination by means of the $\times$ operator of all the preference levels given by the constraints to the corresponding subtuples of the solution, or more formally,

$$\Pi_{c \in C} \text{def}_c(s \downarrow_{\text{con}_c}),$$

where Π is the multiplicative operator of the semiring and $\text{def}_c(s \downarrow_{\text{con}_c})$ is the preference associated by the constraint c to the projection of the solution s on the variables in con_c.

A solution is called **optimal** if there is no other solution with a strictly higher preference.

Three widely used instances of SCSPs are:

- **Classical CSPs** (in short **CSPs**), based on the c-semiring $\langle \{0, 1\}, \vee, \wedge, 0, 1 \rangle$. They model the customary CSPs in which tuples are either allowed or not. So CSPs can be seen as a special case of SCSPs.
- **Fuzzy CSPs**, based on the **fuzzy c-semiring** $\langle [0, 1], max, min, 0, 1 \rangle$. In such problems, preferences are the values in $[0, 1]$, combined by taking the minimum and the goal is to maximize the minimum preference.
- **Weighted CSPs**, based on the **weighted c-semiring** $\langle \Re_+, min, +, \infty, 0 \rangle$. Preferences are costs ranging over non-negative reals, which are aggregated using the sum. The goal is to minimize the total cost.

A simple example of a fuzzy CSP is the following one:

- three variables: x, y, and z, each with the domain $\{a, b\}$;
- two constraints: C_{xy} (over x and y) and C_{yz} (over y and z) defined by:
  $C_{xy} := \{(aa, 0.4), (ab, 0.1), (ba, 0.3), (bb, 0.5)\}$,
  $C_{yz} := \{(aa, 0.4), (ab, 0.3), (ba, 0.1), (bb, 0.5)\}$.

The unique optimal solution of this problem is bbb (an abbreviation for $x = y = z = b$). Its preference is 0.5.

The semiring-based formalism allows one to model also optimization problems with several criteria. This is done by simply considering SCSPs defined on c-semirings which are the Cartesian product of linearly ordered c-semirings. For example, the c-semiring

$$\langle [0, 1] \times [0, 1], (max, max), (min, min), (\mathbf{0}, \mathbf{0}), (\mathbf{1}, \mathbf{1}) \rangle$$

is the Cartesian product of two fuzzy c-semirings. In a SCSP based on such a c-semiring, preferences are pairs, e.g. $(0.1, 0.9)$, combined using the min operator on each component, e.g. $(0.1, 0.8) \times (0.3, 0.6) = (0.1, 0.6)$. The Pareto ordering induced by using the *max* operator on each component is a partial ordering. In this ordering, for example, $(0.1, 0.6) < (0.2, 0.8)$, while $(0.1, 0.9)$ is incomparable to $(0.9, 0.1)$. More generally, if we consider the Cartesian product of n semirings, we end up with a semiring whose elements are tuples of n preferences, each coming from one of the given semirings. Two of such tuples are then ordered if each element in one of them is better or equal to the corresponding one in the other tuple according to the relevant semiring.

2.2   Strategic Games

Let us recall now the notion of a strategic game, see, e.g., [8]. A strategic game for a set N of n players $(n > 1)$ is a sequence

$$(S_1, \ldots, S_n, p_1, \ldots, p_n),$$

where for each $i \in [1..n]$

- S_i is the non-empty set of **strategies** available to player i,
- p_i is the **payoff function** for the player i, so $p_i : S_1 \times \ldots \times S_n \to A$, where A is some fixed linearly ordered set[1].

Given a sequence of non-empty sets $S_1, \ldots, S_n$ and $s \in S_1 \times \ldots \times S_n$ we denote the ith element of s by s_i, abbreviate $N \setminus \{i\}$ to $-i$, and use the following standard notation of game theory, where $i \in [1..n]$ and $I := i_1, \ldots, i_k$ is a subsequence of $1, \ldots, n$:

- $s_I := (s_{i_1}, \ldots, s_{i_k})$,
- $(s'_i, s_{-i}) := (s_1, \ldots, s_{i-1}, s'_i, s_{i+1}, \ldots, s_n)$, where we assume that $s'_i \in S_i$,
- $S_I := S_{i_1} \times \ldots \times S_{i_k}$.

A joint strategy s is called

- a **pure Nash equilibrium** (from now on, simply **Nash equilibrium**) iff

$$p_i(s) \geq p_i(s'_i, s_{-i}) \tag{1}$$

for all $i \in [1..n]$ and all $s'_i \in S_i$,
- **Pareto efficient** if for no joint strategy s', $p_i(s') \geq p_i(s)$ for all $i \in [1..n]$ and $p_i(s') > p_i(s)$ for some $i \in [1..n]$.

Pareto efficiency can be alternatively defined by considering the following strict **Pareto ordering** $<_P$ on the n-tuples of reals:

$$(a_1, \ldots, a_n) <_P (b_1, \ldots, b_n) \text{ iff } \forall i \in [1..n] \; a_i \leq b_i \text{ and } \exists i \in [1..n] \; a_i < b_i.$$

Then a joint strategy s is Pareto efficient iff the n-tuple $(p_1(s), \ldots, p_n(s))$ is a maximal element in the $<_P$ ordering on such n tuples of reals.

To clarify these notions consider the classical Prisoner's Dilemma game represented by the following bimatrix representing the payoffs to both players:

---

[1] The use of A instead of the set of real numbers precludes the construction of mixed strategies and hence of Nash equilibria in mixed strategies, but is sufficient for our purposes.

	C_2	N_2
C_1	3, 3	0, 4
N_1	4, 0	1, 1

Each player i represents a prisoner, who has two strategies, C_i (cooperate) and N_i (not cooperate). Table entries represent payoffs for the players (where the first component is the payoff of player 1 and the second one that of player 2).

The two prisoners gain when both cooperate (a gain of 3 each). However, if only one of them cooperates, the other one, who does not cooperate, will gain more (a gain of 4). If both do not cooperate, both gain very little (that is, 1 each), but more than the "cheated" cooperator whose cooperation is not returned (that is, 0).

Here the unique Nash equilibrium is (N_1, N_2), while the other three joint strategies (C_1, C_2), (C_1, N_2) and (N_1, C_2) are Pareto efficient.

2.3   Graphical Games

A related modification of the concept of strategic games, called **graphical games**, was proposed in [7]. These games stress the locality in taking decision. In a graphical game the payoff of each player depends only on the strategies of its neighbours in a given in advance graph structure over the set of players.

More formally, a **graphical game** for n players with the corresponding strategy sets $S_1, \ldots, S_n$ with the payoffs being elements of a linearly ordered set A, is defined by assuming a neighbour function $neigh$ that given a player i yields its set of neighbours $neigh(i)$. The payoff for player i is then a function p_i from $\Pi_{j \in neigh(i) \cup \{i\}} S_j$ to A. We denote such a graphical game by

$$(S_1, \ldots, S_n, neigh, p_1, \ldots, p_n, A).$$

By using the canonical extensions of these payoff functions to the Cartesian product of all strategy sets one can then extend the previously introduced concepts to the graphical games. Further, when all pairs of players are neighbours, a graphical game reduces to a strategic game.

3   Optimality in SCSPs and Nash Equilibria in Graphical Games

In this section we relate the notion of optimality in soft constraints and the concept of Nash equilibria in graphical games. We shall see that, while CSPs are sufficient to obtain the Nash equilibria of any given graphical game, the opposite direction does not hold. However, graphical games can model, via their Nash equilibria, a superset of the set of the optimal solutions of any given SCSP.

The first statement is based on a result in [5], where, given a graphical game, it is shown how to build a corresponding CSP such that the Nash equilibria of the game and the solutions of the CSP coincide. Thus, the full expressive power of SCSPs is not needed to model the Nash equilibria of a game. We will now focus on the opposite direction: from SCSPs to graphical games. Unfortunately, the inverse of the mapping defined in [5] cannot be used for this purpose since it only returns CSPs of a specific kind.

3.1   From SCSPs to Graphical Games: A Local Mapping

We now define a mapping from soft CSPs to a specific kind of graphical games. We identify the players with the variables. Thus, since soft constraints link variables, the resulting game players are naturally connected. To capture this aspect, we use graphical games. We allow here payoffs to be elements of an arbitrary linearly ordered set.

Let us consider a first possible mapping from SCSPs to graphical games. In what follows we focus on SCSPs based on c-semirings with the carrier linearly ordered by $\leq$ (e.g. fuzzy or weighted) and on the concepts of optimal solutions in SCSPs and Nash equilibria in graphical games.

Given a SCSP $P := \langle C, V, D, S \rangle$ we define the corresponding graphical game for $n = |V|$ players as follows:

- the players: one for each variable;
- the strategies of player i: all values in the domain of the corresponding variable x_i;
- the neighbourhood relation: $j \in neigh(i)$ iff the variables x_i and x_j appear together in some constraint from C;
- the payoff function of player i:
  Let $C_i \subseteq C$ be the set of constraints involving x_i and let X be the set of variables that appear together with x_i in some constraint in C_i (i.e., $X = \{x_j \mid j \in neigh(i)\}$). Then given an assignment s to all variables in $X \cup \{x_i\}$ the payoff of player i w.r.t. s is defined by:

$$p_i(s) := \Pi_{c \in C_i} \mathrm{def}_c(s \downarrow_{\mathrm{con}_c}).$$

We denote the resulting graphical game by $L(P)$ to emphasize the fact that the payoffs are obtained using *local* information about each variable, by looking only at the constraints in which it is involved.

One could think of a different mapping where players correspond to constraints. However, such a mapping can be obtained by applying the local mapping L to the hidden variable encoding [13] of the SCSP in input.

We now analyze the relation between the optimal solutions of a SCSP P and the Nash equilibria of the derived game $L(P)$.

3.1.1   General Case

In general, these two concepts are unrelated. Indeed, consider the fuzzy CSP defined at the end of Section 2.1. The corresponding game has:

- three players, x, y, and z;
- each player has two strategies, a and b;

- the neighbourhood relation is defined by:

$$neigh(x) := \{y\}, \ neigh(y) := \{x, z\}, \ neigh(z) := \{y\};$$

- the payoffs of the players are defined as follows:
  - for player x:
    $p_x(aa*) := 0.4, \ p_x(ab*) := 0.1, \ p_x(ba*) := 0.3, \ p_x(bb*) := 0.5;$
  - for player y:
    $p_y(aaa) := 0.4, \ p_y(aab) := 0.3, \ p_y(abb) := 0.1, \ p_y(bbb) := 0.5,$
    $p_y(bba) := 0.5, \ p_y(baa) := 0.3, \ p_y(bab) := 0.3, \ p_y(aba) := 0.1;$
  - for player z:
    $p_z(*aa) := 0.4, \ p_z(*ab) := 0.3, \ p_z(*ba) := 0.1, \ p_z(*bb) := 0.5;$

where $*$ stands for either a or b and where to facilitate the analysis we use the canonical extensions of the payoff functions p_x and p_z to the functions on $\{a, b\}^3$.

This game has two Nash equilibria: aaa and bbb. However, only bbb is an optimal solution of the fuzzy SCSP.

One could thus think that in general the set of Nash equilibria is a superset of the set of optimal solutions of the corresponding SCSP. However, this is not the case. Indeed, consider a fuzzy CSP with as before three variables, x, y and z, each with the domain $\{a, b\}$, but now with the constraints:

$C_{xy} := \{(aa, 0.9), (ab, 0.6), (ba, 0.6), (bb, 0.9)\},$
$C_{yz} := \{(aa, 0.1), (ab, 0.2), (ba, 0.1), (bb, 0.2)\}.$

Then aab, abb, bab and bbb are all optimal solutions but only aab and bbb are Nash equilibria of the corresponding graphical game.

3.1.2   SCSPs with Strictly Monotonic Combination

Next, we consider the case when the multiplicative operator $\times$ is strictly monotonic. Recall that given a c-semiring $\langle A, +, \times, \mathbf{0}, \mathbf{1} \rangle$, the operator $\times$ is **strictly monotonic** if for any $a, b, c \in A$ such that $a < b$ we have $c \times a < c \times b$. (The symmetric condition is taken care of by the commutativity of $\times$.)

Note for example that in the case of classical CSPs $\times$ is not strictly monotonic, as $a < b$ implies that $a = 0$ and $b = 1$ but $c \wedge a < c \wedge b$ does not hold then for $c = 0$. Also in fuzzy CSPs $\times$ is not strictly monotonic, as $a < b$ does not imply that $min(a, c) < min(b, c)$ for all c. In contrast, in weighted CSP $\times$ is strictly monotonic, as $a < b$ in the carrier means that $b < a$ as reals, so for any c we have $c + b < c + a$, i.e., $c \times a < c \times b$ in the carrier.

So consider now a c-semiring with a linearly ordered carrier and a strictly monotonic multiplicative operator. As in the previous case, given an SCSP P, it is possible that a Nash equilibrium of $L(P)$ is not an optimal solution of P. Consider for example a weighted SCSP P with

- two variables, x and y, each with the domain $D = \{a, b\}$;
- one constraint $C_{xy} := \{(aa, 3), (ab, 10), (ba, 10), (bb, 1)\}$.

The corresponding game $L(P)$ has:

- two players, x and y, who are neighbours of each other;
- each player has two strategies, a and b;
- the payoffs defined by:
  $p_x(aa) := p_y(aa) := 7$, $p_x(ab) := p_y(ab) := 0$,
  $p_x(ba) := p_y(ba) := 0$, $p_x(bb) := p_y(bb) := 9$.

Notice that, in a weighted CSP we have $a \leq b$ in the carrier iff $b \leq a$ as reals, so when passing from the SCSP to the corresponding game, we have complemented the costs w.r.t. 10, when making them payoffs. In general, given a weighted CSP, we can define the payoffs (which must be maximized) from the costs (which must be minimized) by complementing the costs w.r.t. the greatest cost used in any constraint of the problem.

Here $L(P)$ has two Nash equilibria, aa and bb, but only bb is an optimal solution. Thus, as in the fuzzy case, we have that there can be a Nash equilibrium of $L(P)$ that is not an optimal solution of P. However, in contrast to the fuzzy case, when the multiplicative operator of the SCSP is strictly monotonic, the set of Nash equilibria of $L(P)$ is a superset of the set of optimal solutions of P.

Theorem 1. *Consider a SCSP P defined on a c-semiring $\langle A, +, \times, \mathbf{0}, \mathbf{1} \rangle$, where A is linearly ordered and $\times$ is strictly monotonic, and the corresponding game $L(P)$. Then every optimal solution of P is a Nash equilibrium of $L(P)$.*

Proof. We prove that if a joint strategy s is not a Nash equilibrium of game $L(P)$, then it is not an optimal solution of SCSP P.

Let a be the strategy of player x in s, and let $s_{neigh(x)}$ and s_Y be, respectively, the joint strategy of the neighbours of x, and of all other players, in s. That is, $V = \{x\} \cup neigh(x) \cup Y$ and we write s as $(a, s_{neigh(x)}, s_Y)$.

By assumption there is a strategy b for x such that the payoff $p_x(s')$ for the joint strategy $s' := (b, s_{neigh(x)}, s_Y)$ is higher than $p_x(s)$. (We use here the canonical extension of p_x to the Cartesian product of all the strategy sets).

So by the definition of the mapping L

$$\Pi_{c \in C_x} \mathrm{def}_c(s \downarrow_{\mathrm{con}_c}) < \Pi_{c \in C_x} \mathrm{def}_c(s' \downarrow_{\mathrm{con}_c}),$$

where C_x is the set of all the constraints involving x in SCSP P. But the preference of s and s' is the same on all the constraints not involving x and $\times$ is strictly monotonic, so we conclude that

$$\Pi_{c \in C} \mathrm{def}_c(s \downarrow_{\mathrm{con}_c}) < \Pi_{c \in C} \mathrm{def}_c(s' \downarrow_{\mathrm{con}_c}).$$

This means that s is not an optimal solution of P. $\square$

3.1.3 Classical CSPs

The above result does not hold for classical CSPs. Indeed, consider a CSP with:

- three variables: x, y, and z, each with the domain $\{a, b\}$;
- two constraints: C_{xy} (over x and y) and C_{yz} (over y and z) defined by:
  $C_{xy} := \{(aa, 1), (ab, 0), (ba, 0), (bb, 0)\}$,
  $C_{yz} := \{(aa, 0), (ab, 0), (ba, 1), (bb, 0)\}$.

This CSP has no solutions in the classical sense, i.e., each optimal solution, in particular baa, has preference 0. However, baa is not a Nash equilibrium of the resulting graphical game, since the payoff of player x increases when he switches to the strategy a.

On the other hand, if we restrict the domain of L to consistent CSPs, that is, CSPs with at least one solution with value 1, then the discussed inclusion does hold.

Proposition 1. *Consider a consistent CSP P and the corresponding game $L(P)$. Then every solution of P is a Nash equilibrium of $L(P)$.*

Proof. Consider a solution s of P. In the resulting game $L(P)$ the payoff to each player is maximal, namely 1. So the joint strategy s is a Nash equilibrium in game $L(P)$. □

The reverse inclusion does not need to hold. Indeed, consider the following CSP:

- three variables: x, y, and z, each with the domain $\{a, b\}$;
- two constraints: C_{xy} and C_{yz} defined by:
  $C_{xy} := \{(aa, 1), (ab, 0), (ba, 0), (bb, 0)\}$,
  $C_{yz} := \{(aa, 1), (ab, 0), (ba, 0), (bb, 0)\}$.

Then aaa is a solution, so the CSP is consistent. But bbb is not an optimal solution, while it is a Nash equilibrium of the resulting game.

So for consistent CSPs our mapping L yields games in which the set of Nash equilibria is a, possibly strict, superset of the set of solutions of the CSP.

However, there are ways to relate CSPs and games so that the solutions and the Nash equilibria coincide. This is what is done in [5], where the mapping is from the strategic games to CSPs. Notice that our mapping goes in the opposite direction and it is not the reverse of the one in [5]. In fact, the mapping in [5] is not reversible.

3.2   From SCSPs to Graphical Games: A Global Mapping

Other mappings from SCSPs to games can be defined. While our mapping L is in some sense 'local', since it considers the neighbourhood of each variable, we can also define an alternative 'global' mapping that considers all constraints. More precisely, given a SCSP $P = \langle C, V, D, S \rangle$, with a linearly ordered carrier A of S, we define the corresponding game on $n = |V|$ players, $GL(P) = (S_1, \ldots, S_n, p_1, \ldots, p_n, A)$ by using the following payoff function p_i for player i:

- given an assignment s to *all* variables in V

$$p_i(s) := \Pi_{c \in C} \mathrm{def}_c(s \downarrow_{\mathrm{con}_c}).$$

Notice that in the resulting game the payoff functions of all players are the same.

Theorem 2. *Consider an SCSP P over a linearly ordered carrier, and the corresponding game GL(P). Then every optimal solution of P is a Nash equilibrium of GL(P).*

Proof. An optimal solution of P, say s, is a joint strategy for which all players have the same, highest, payoff. So no other joint strategy exists for which some player is better off and consequently s is a Nash equilibrium.   □

The opposite inclusion does not need to hold. Indeed, consider again the weighted SCSP of Subsection 3.1 with

- two variables, x and y, each with the domain $D = \{a, b\}$;
- one constraint, $C_{xy} := \{(aa, 3), (ab, 10), (ba, 10), (bb, 1)\}$.

Since there is one constraint, the mappings L and GL coincide. Thus we have that aa is a Nash equilibrium of $GL(P)$ but is not an optimal solution of P.

While the mapping defined in this section has the advantage of providing a precise subset relationship between optimal solutions and Nash equilibria, as Theorem 2 states, it has an obvious disadvantage from the computational point of view, since it requires to consider all the complete assignments of the SCSP.

3.3   Summary of Results

Summarizing, in this section we have analyzed the relationship between the optimal solutions of SCSPs and the Nash equilibria of graphical games. In [5] CSPs have been shown to be sufficient to model Nash equilibria of graphical games. Here we have considered the question whether the Nash equilibria of graphical games can model the optimal solutions of SCSPs. We have provided two mappings from SCSPs to graphical games, showing that (with some conditions for the local mapping) the set of Nash equilibria of the obtained game contains the optimal solutions of the given SCSP.

Nash equilibria can be seen as the optimal elements in very specific orderings, where dominance is based on exactly one change in the joint strategy, while SCSPs can model any ordering. So we conjecture that it is not possible to find a mapping from SCSPs to the graphical games for which the optimals coincide with Nash equilibria. Such a conjecture is also supported by the fact that strict Nash equilibria can be shown to coincide with the optimals of a CP-net, see [1], and the CP-nets can model strictly less orderings than the SCSPs, see [10].

4   Optimality in SCSPs and Pareto Efficient Joint Strategies in Graphical Games

Next, we relate the notion of optimality in SCSPs to the Pareto efficient joint strategies of graphical games.

4.1   From SCSPs to Graphical Games

Consider again the local and the global mappings from SCSPs to graphical games defined in Sections 3.1 and 3.2. We will now prove that the local mapping yields a game whose set of Pareto efficient joint strategies contains the set of optimal solutions of a given SCSP. On the other hand, the global mapping gives a one-to-one correspondence between the two sets.

Theorem 3. *Consider an SCSP P defined on a c-semiring $\langle A, +, \times, \mathbf{0}, \mathbf{1} \rangle$, where A is linearly ordered and $\times$ is strictly monotonic, and the corresponding game $L(P)$. Then every optimal solution of P is a Pareto efficient joint strategy of $L(P)$.*

Proof. Let us consider a joint strategy s of L(P) which is not Pareto efficient. We will show that s does not correspond to an optimal solution of P. Since s is not Pareto efficient, there is a joint strategy s' such that $p_i(s) \leq p_i(s')$ for all $i \in [1..n]$ and $p_i(s) < p_i(s')$ for some $i \in [1..n]$. Let us denote with $I = \{i \in [1..n]$ such that $p_i(s) < p_i(s')\}$. By the definition of the mapping L, we have:

$$\Pi_{c \in C_i} \mathrm{def}_c(s \downarrow_{\mathrm{con}_c}) < \Pi_{c \in C_i} \mathrm{def}_c(s' \downarrow_{\mathrm{con}_c}),$$

for all $i \in I$ and where C_i is the set of all the constraints involving the variable corresponding to player i in SCSP P. Since the preference of s and s' is the same on all the constraints not involving any $i \in I$, and since $\times$ is strictly monotonic, we have:

$$\Pi_{c \in C} \mathrm{def}_c(s \downarrow_{\mathrm{con}_c}) < \Pi_{c \in C} \mathrm{def}_c(s' \downarrow_{\mathrm{con}_c}).$$

This means that s is not an optimal solution of P.   □

To see that there may be Pareto efficient joint strategies that do not correspond to the optimal solutions, consider a weighted SCSP P with

- two variables, x and y, each with domain $D = \{a, b\}$;
- constraint $C_x := \{(a, 2), (b, 1)\}$;
- constraint $C_y := \{(a, 4), (b, 7)\}$;
- constraint $C_{xy} := \{(aa, 0), (ab, 10), (ba, 10), (bb, 0)\}$.

The corresponding game $L(P)$ has:

- two players, x and y, who are neighbours of each other;
- each player has two strategies: a and b;
- the payoffs defined by: $p_x(aa) := 8$, $p_y(aa) := 6$, $p_x(ab) := p_y(ab) := 0$, $p_x(ba) := p_y(ba) := 0$, $p_x(bb) := 9$, $p_y(bb) := 3$.

As in Section 3.1 when passing from an SCSP to the corresponding game, we have complemented the costs w.r.t. 10, when turning them to payoffs. $L(P)$ has two Pareto efficient joint strategies: aa and bb. (They are also both Nash equilibria.) However, only aa is optimal in P.

If the combination operator is idempotent, there is no relation between the optimal solutions of P and the Pareto efficient joint strategies of $L(P)$. However, if we use the global mapping defined in Section 3.2, the optimal solutions do correspond to Pareto efficient joint strategies, regardless of the type of the combination operator.

Theorem 4. *Consider an SCSP P defined on a c-semiring $\langle A, +, \times, \mathbf{0}, \mathbf{1} \rangle$, where A is linearly ordered, and the corresponding game $GL(P)$. Then every optimal solution of P is a Pareto efficient joint strategy of $GL(P)$, and viceversa.*

Proof. Any optimal solution corresponds to a joint strategy where all players have the same payoff, which is the solution's preference. Thus, such a joint strategy cannot be Pareto dominated by any other strategy. Conversely, a solution corresponding to a joint strategy with the highest payoff is optimal.   □

4.2   From Graphical Games to SCSPs

Next, we define a mapping from graphical games to SCSPs that relates Pareto efficient joint strategies in games to optimal solutions in SCSPs. In order to define such a mapping, we limit ourselves to SCSPs defined on c-semirings which are the Cartesian product of linearly ordered c-semirings (see Section 2.1). More precisely, given a graphical game $G = (S_1, \ldots, S_n, neigh, p_1, \ldots, p_n, A)$ we define the corresponding SCSP $L'(G) = \langle C, V, D, S \rangle$, as follows:

- each variable x_i corresponds to a player i;
- the domain $D(x_i)$ of the variable x_i consists of the set of strategies of player i, i.e., $D(x_i) := S_i$;
- the c-semiring is
  $$\langle A_1 \times \cdots \times A_n, (+_1, \ldots, +_n), (\times_1, \ldots, \times_n), (\mathbf{0}_1, \ldots, \mathbf{0}_n), (\mathbf{1}_1, \ldots, \mathbf{1}_n) \rangle,$$
  the Cartesian product of n *arbitrary* linearly ordered semirings;
- soft constraints: for each variable x_i, one constraint $\langle def, con \rangle$ such that:
  - $con = neigh(x_i) \cup \{x_i\}$;
  - $def : \Pi_{y \in con} D(y) \to A_1 \times \cdots \times A_n$ such that for any $s \in \Pi_{y \in con} D(y)$, $def(s) := (d_1, \ldots, d_n)$ with $d_j = \mathbf{1}_j$ for every $j \neq i$ and $d_i = f(p_i(s))$, where $f : A \to A_i$ is an order preserving mapping from payoffs to preferences (i.e., if $r > r'$ then $f(r) > f(r')$ in the c-semiring's ordering).

To illustrate it consider again the previously used Prisoner's Dilemma game:

	C_2	N_2
C_1	3, 3	0, 4
N_1	4, 0	1, 1

Recall that in this game the only Nash equilibrium is (N_1, N_2), while the other three joint strategies are Pareto efficient.

We shall now construct a corresponding SCSP based on the Cartesian product of two weighted semirings. This SCSP according to the mapping L' has:[2]

---

[2] Recall that in the weighted semiring **1** equals 0.

- two variables: x_1 and x_2, each with the domain $\{c, n\}$;
- two constraints, both on x_1 and x_2:
  - constraint c_1 with $\operatorname{def}(cc) := \langle 7, 0 \rangle$, $\operatorname{def}(cn) := \langle 10, 0 \rangle$, $\operatorname{def}(nc) := \langle 6, 0 \rangle$, $\operatorname{def}(nn) := \langle 9, 0 \rangle$;
  - constraint c_2 with $\operatorname{def}(cc) := \langle 0, 7 \rangle$, $\operatorname{def}(cn) := \langle 0, 6 \rangle$, $\operatorname{def}(nc) := \langle 0, 10 \rangle$, $\operatorname{def}(nn) := \langle 0, 9 \rangle$;

The optimal solutions of this SCSPs are: cc, with preference $\langle 7, 7 \rangle$, nc, with preference $\langle 10, 6 \rangle$, cn, with preference $\langle 6, 10 \rangle$. The remaining solution, nn, has a lower preference in the Pareto ordering. Indeed, its preference $\langle 9, 9 \rangle$ is dominated by $\langle 7, 7 \rangle$, the preference of cc (since preferences are here costs and have to be minimized). Thus the optimal solutions coincide here with the Pareto efficient joint strategies of the given game. This is true in general.

Theorem 5. *Consider a graphical game G and a corresponding SCSP $L'(G)$. Then the optimal solutions of $L'(G)$ coincide with the Pareto efficient joint strategies of G.*

Proof. In the definition of the mapping L' we stipulated that the mapping f maintains the ordering from the payoffs to preferences. As a result each joint strategy s corresponds to the n-tuple of preferences $(f(p_1(s)), \ldots, f(p_n(s)))$ and the Pareto orderings on the n-tuples $(p_1(s), \ldots, p_n(s))$ and $(f(p_1(s)), \ldots, f(p_n(s)))$ coincide. Consequently a sequence s is an optimal solution of the SCSP $L'(G)$ iff $(f(p_1(s)), \ldots, f(p_n(s)))$ is a maximal element of the corresponding Pareto ordering. □

We notice that L' is injective and, thus, can be reversed on its image. When such a reverse mapping is applied to these specific SCSPs, payoffs correspond to projecting of the players' valuations to a subcomponent.

4.2.1   Pareto Efficient Nash Equilibria

As mentioned earlier, in [5] a mapping is defined from the graphical games to CSPs such that Nash equilibria coincide with the solutions of CSP. Instead, our mapping is from the graphical games to SCSPs, and is such that Pareto efficient joint strategies and the optimal solutions coincide.

Since CSPs can be seen as a special instance of SCSPs, where only **1**, **0**, the top and bottom elements of the semiring, are used, it is possible to add to any SCSP a set of hard constraints. Therefore we can merge the results of the two mappings into a single SCSP, which contains the soft constraints generated by L' and also the hard constraints generated by the mapping in [5], Below we denote these hard constraints by $H(G)$. We recall that each constraint in $H(G)$ corresponds to a player, has the variables corresponding to the player and it neighbours and allows only tuples corresponding to the strategies in which the player has no so-called regrets. If we do this, then the optimal solutions of the new SCSP with preference higher than **0** are the Pareto efficient Nash equilibria of the given game, that is, those Nash equilibria which dominate or are incomparable with all other Nash equilibria according to the Pareto ordering. Formally, we have the following result.

Theorem 6. *Consider a graphical game G and the SCSP $L'(G) \cup H(G)$. If the optimal solutions of $L'(G) \cup H(G)$ have global preference greater than $\mathbf{0}$, they correspond to the Pareto efficient Nash equilibria of G.*

Proof. Given any solution s, let p be its preference in $L'(G)$ and p' in $L'(G) \cup H(G)$. By the construction of the constraints $H(G)$ we have that p' equals p if s is a Nash equilibrium and p' equals $\mathbf{0}$ otherwise. The remainder of the argument is as in the proof of Theorem 5. $\qquad\Box$

For example, in the Prisoner's Dilemma game, the mapping in [5] would generate just one constraint on x_1 and x_2 with nn as the only allowed tuple. In our setting, when using as the linearly ordered c-semirings the weighted semirings, this would become a soft constraint with

$$\mathrm{def}(cc) := \mathrm{def}(cn) := \mathrm{def}(nc) = \langle \infty, \infty \rangle, \ \mathrm{def}(nn) := \langle 0, 0 \rangle.$$

With this new constraint, all solutions have the preference $\langle \infty, \infty \rangle$, except for nn which has the preference $\langle 9, 9 \rangle$ and thus is optimal. This solution corresponds to the joint strategy (N_1, N_2) with the payoff $(1, 1)$ (and thus preference $(9, 9)$). This is the only Nash equilibrium and thus the only Pareto efficient Nash equilibrium.

This method allows us to identify among Nash equilibria the 'optimal' ones. One may also be interested in knowing whether there exist Nash equilibria which are also Pareto efficient joint strategies. For example, in the Prisoners' Dilemma example, there are no such Nash equilibria. To find any such joint strategies we can use the two mappings separately, to obtain, given a game G, both an SCSP $L'(G)$ and a CSP $H(G)$ (using the mapping in [5]). Then we should take the intersection of the set of optimal solutions of $L'(G)$ and the set of solutions of $H(G)$.

4.3   Summary of Results

We have considered the relationship between optimal solutions of SCSPs and Pareto efficient joint strategies in graphical games. The local mapping of Section 3.1 turns out to map optimal solutions of a given SCSP to Pareto efficient joint strategies, while the global mapping of Section 3.2 yields a one-to-one correspondence. For the reverse direction it is possible to define a mapping such that these two notions of optimality coincide. However, none of these mappings are onto.

5   Conclusions

In this paper we related two formalisms that are commonly used to reason about optimal outcomes: graphical games and soft constraints. While for soft constraints there is only one notion of optimality, for graphical games there are at least two. In this paper we have considered Nash equilibria and Pareto efficient joint strategies.

We have defined a natural mapping from SCSPs that combine preferences using a strictly monotonic operator to a class of graphical games such that the optimal solutions of the SCSP are included in the Nash equilibria of the game and in the set of Pareto efficient joint strategies. In general the inclusions cannot be reversed. We have also exhibited a mapping from the graphical games to a class of SCSPs such that the Pareto efficient joint strategies of the game coincide with the optimal solutions of the SCSP.

These results can be used in many ways. One obvious way is to try to exploit computational and algorithmic results existing for one of these areas in another. This has been pursued already in [5] for games by using hard constraints. As a consequence of our results this can also be done for strategic games by using soft constraints. For example, finding a Pareto efficient joint strategy involves mapping a game into an SCSP and then solving it. A similar approach can also be applied to Pareto efficient Nash equilibria, which can be found by solving a suitable SCSP.

References

[1] Apt, K.R., Rossi, F., Venable, K.B.: CP-nets and Nash equilibria. In: Proc. of the Third International Conference on Computational Intelligence, Robotics and Autonomous Systems (CIRAS 2005), pp. 1–6 (2005), http://arxiv.org/abs/cs/0509071

[2] Bistarelli, S., Montanari, U., Rossi, F.: Semiring-based constraint solving and optimization. Journal of the ACM 44(2), 201–236 (1997)

[3] Boutilier, C., Brafman, R.I., Domshlak, C., Hoos, H.H., Poole, D.: CP-nets: A tool for representing and reasoning with conditional ceteris paribus preference statements. J. Artif. Intell. Res (JAIR) 21, 135–191 (2004)

[4] Fargier, H., Dubois, D., Prade, H.: The calculus of fuzzy restrictions as a basis for flexible constraint satisfaction. In: IEEE International Conference on Fuzzy Systems (1993)

[5] Greco, G., Gottlob, G., Scarcello, F.: Pure Nash equilibria: hard and easy games. J. of Artificial Intelligence Research 24, 357–406 (2005)

[6] Greco, G., Scarcello, F.: Constrained pure Nash equilibria in graphical games. In: Proceedings of the 16th Eureopean Conference on Artificial Intelligence (ECAI 2004), pp. 181–185. IOS Press, Amsterdam (2004)

[7] Kearns, M., Littman, M., Singh, S.: Graphical models for game theory. In: Proceedings of the 17th Conference in Uncertainty in Artificial Intelligence (UAI 2001), pp. 253–260. Morgan Kaufmann, San Francisco (2001)

[8] Myerson, R.B.: Game Theory: Analysis of Conflict. Harvard Univ. Press, Cambridge (1991)

[9] Rossi, F., Meseguer, P., Schiex, T.: Soft constraints. In: Walsh, T., Rossi, F., Van Beek, P. (eds.) Handbook of Constraint programming, pp. 281–328. Elsevier, Amsterdam (2006)

[10] Domshlak, C., Prestwich, S., Rossi, F., Venable, K.B., Walsh, T.: Hard and soft constraints for reasoning about qualitative conditional preferences. Journal of Heuristics, Special issue on preferences 12, 263–285 (2006)

[11] Ruttkay, Z.: Fuzzy constraint satisfaction. In: Proceedings 1st IEEE Conference on Evolutionary Computing, Orlando, pp. 542–547 (1994)

[12] Maheswaran, R.T., Pearce, J.P., Tambe, M.: Distributed algorithms for DCOP: a graphical-game-based approach. In: Proceedings of the ISCA 17th International Conference on Parallel and Distributed Computing Systems (ISCA PDCS 2004), pp. 432–439. ISCA (2004)

[13] Mamoulis, N., Stergiou, K.: Solving non-binary CSPs using the hidden variable encoding. In: Walsh, T. (ed.) CP 2001. LNCS, vol. 2239. Springer, Heidelberg (2001)

Temporal Reasoning in Nested Temporal Networks with Alternatives

Roman Barták, Ondřej Čepek, and Martin Hejna

Charles University in Prague, Faculty of Mathematics and Physics
Malostranské nám. 2/25, 118 00 Praha 1, Czech Republic
roman.bartak@mff.cuni.cz, ondrej.cepek@mff.cuni.cz,
mhejna@matfyz.cz

Abstract. Temporal networks play a crucial role in modeling temporal relations in planning and scheduling applications. Temporal Networks with Alternatives (TNAs) were proposed to model alternative and parallel processes in production scheduling, however the problem of deciding which nodes can be consistently included in such networks is NP-complete. A tractable subclass, called Nested TNAs, can still cover a wide range of real-life processes, while the problem of deciding node validity is solvable in polynomial time. In this paper, we show that adding simple temporal constraints (instead of precedence relations) to Nested TNAs makes the problem NP-hard again. We also present several complete and incomplete techniques for temporal reasoning in Nested TNAs.

1 Introduction

Planning and scheduling applications almost always include some form of temporal reasoning, for example, a causal relation (the effect of some activity is required for processing another activity) implies a precedence constraint. These relations are frequently modeled using temporal networks where nodes correspond to activities and arcs are annotated by the temporal relations between activities. Current temporal networks handle well temporal information including disjunction of temporal constraints [13] or uncertainty [4]. Several other extensions of temporal networks appeared recently such as resource temporal networks [10] or disjunctive temporal networks with finite domain constraints [11]. These extensions integrate temporal reasoning with reasoning on non-temporal information, such as fluent resources (for example fuel consumption during car driving). All these approaches assume that all nodes are present in the network, though the position of nodes in time may be influenced by other than temporal constraints. Conditional Temporal Planning [14] introduced an option to decide which node will be present in the solution depending on a certain external condition. Hence CTP can model conditional plans where the nodes actually present in the solution are selected based on external forces. Temporal Plan Networks [8] (TPN) also include conditional branching and they attempt to model all alternative plans in a single graph. Temporal Networks with Alternatives [1] (TNA) introduced a different type of alternatives with so called parallel and alternative branching. They are more general than TPN but the problem of deciding which nodes can be

F. Fages, F. Rossi, and S. Soliman (Eds.): CSCLP 2007, LNAI 5129, pp. 17–31, 2008.

consistently included in the network, if some nodes are pre-selected, is NP-complete even if no temporal constraints are imposed. Therefore a restricted form, so called Nested TNAs, was proposed in [2]. Nested TNAs have a similar topology as TPNs though the original motivation for their introduction was different – a Nested TNA focuses on manufacturing processes while a TPN models plans for unmanned vehicles. The paper [2] shows that the problem of deciding whether a subset of nodes can be selected to satisfy the branching constraints is now tractable, but it still leaves open the question what happens if temporal constraints are assumed. In this paper we present a new complexity result for Nested Temporal Networks with Alternatives where simple temporal constraints are included. We also present some new algorithms that can help in solving problems based on (Nested) TNAs. These algorithms exploit the integrated reasoning on both logical (branching) and temporal constraints.

There exist other frameworks mixing temporal and logical reasoning. In problems, such as log-based reconciliation [7], we need to model inter-dependencies between nodes which concern their presence/absence in the final solution. The possibility to select nodes according to logical, temporal, and resource constrains was introduced to manufacturing scheduling by ILOG in their MaScLib [12]. The same idea was independently formalized in Extended Resource Constrained Project Scheduling Problem [9]. In the common model each node has a Boolean validity variable indicating whether the node is selected to be in the solution. These variables are a discrete version of PEX variables used by Beck and Fox [3] for modeling presence of alternative activities in the schedule. In many recent approaches, these variables are interconnected by logical constraints such as the dependency constraint described above.

In this paper, we first give motivation for using Temporal Networks with Alternatives and formally introduce TNAs and their nested form. The main part of the paper shows that using temporal constraints in Nested TNAs makes the problem of deciding which nodes can be consistency included in the network NP-complete again. We also present several techniques that can help in solving the problem. These techniques are proposed in the context of constraint satisfaction so they can be easily integrated with other constraints, for example with constraints that model resources. Hence the proposed techniques are useful for solving oversubscribed real-life scheduling problems.

2 Motivation and Background

Let us consider a manufacturing scheduling problem of piston production. Each piston consists of a rod and a tube that need to be assembled together to form the piston. Each rod consists of the main body and a special kit that is welded to the rod (the kit needs to be assembled before welding). The rod body is sawn from a large metal stick. The tube can also be sawn from a larger tube. Rod body, the kit, and tube must be collected together from the warehouse to ensure that their diameters fit. If the tube is not available, it can be bought from an external supplier. In any case some welding is necessary to be done on the tube before it can be assembled with the rod. Finally, between sawing and welding, both rod and tube must be cleared of metal cuts produced by sawing. Assume that welding and sawing operations require ten time units, assembly operation requires five time units, clearing can be done in two time units, and the material is collected from warehouse in one time unit. If the tube is bought from an external supplier then it takes fifty time units to get it. Moreover, tube and rod must cool-down after welding which takes five time units.

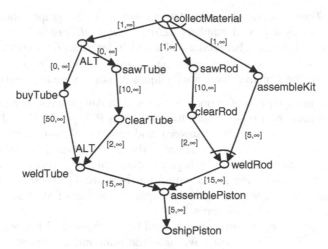

Fig. 1. Example of a manufacturing process with alternatives

The manufacturing processes from the above problem can be described using a Temporal Network with Alternatives depicted in Figure 1. Nodes correspond to start times of operations and arcs are annotated by simple temporal constraints in the form [a, b], where a describes the minimal distance (in time) between the nodes and b describes the maximal distance. Informally, this network describes the traditional simple temporal constraints [5] together with the specification of branching of processes. There is a *parallel branching* marked by a semi-circle indicating that the process splits and runs in parallel and an *alternative branching* marked by ALT indicating that the process will consists of exactly one alternative path (we can choose between buying a tube and producing it in situ).

3 Temporal Networks with Alternatives

Let us now formally define Temporal Networks with Alternatives from [1]. Let G be a directed acyclic graph. A sub-graph of G is called a *fan-out sub-graph* if it consists of nodes $x, y_1,..., y_k$ (for some k) such that each (x, y_i), $1 \le i \le k$, is an arc in G. If $y_1,..., y_k$ are all and the only successors of x in G (there is no z such that (x, z) is an arc in G and $\forall i = 1,...,k: z \ne y_i$) then we call the fan-out sub-graph complete. Similarly, a sub-graph of G is called a *fan-in sub-graph* if it consists of nodes $x, y_1,..., y_k$ (for some k) such that each (y_i, x), $1 \le i \le k$, is an arc in G. A complete fan-in sub-graph is defined similarly as above. In both cases x is called a *principal node* and all $y_1,..., y_k$ are called *branching nodes*.

Definition 1: A directed acyclic graph G together with its pair wise edge-disjoint decomposition into complete fan-out and fan in sub-graphs, where each sub-graph in the decomposition is marked either as a *parallel* sub-graph or an *alternative* sub-graph, is called a *P/A graph*.

Definition 2: *Temporal Network with Alternatives* is a P/A graph where each arc (x, y) is annotated by a pair of numbers $[a,b]$ (a *temporal annotation*) where a describes the minimal distance between x and y and b describes the maximal distance, formally, $a \leq t_y - t_x \leq b$, where t_x denotes the position of node x in time. Frequently, both numbers are non-negative, but our techniques do not require this restriction.

Figure 1 shows an example of Temporal Network with Alternatives. If we remove the temporal constraints from this network then we get a P/A graph. Note that the arcs (*sawTube, clearTube*), (*sawRode, clearRod*), and (*assemblePiston, shipPiston*) form simple fan-in (or fan-out, it does not matter in this case) sub-graphs. As we will see later, it does not matter whether the sub-graphs consisting of a single arc are marked as parallel or alternative – the logical constraint imposed by the sub-graph will be always the same. Hence, we can omit the explicit marking of such single-arc sub-graphs to make the figure less overcrowded.

We call the special logical relations imposed by the fan-in and fan-out sub-graphs *branching constraints*. Temporarily, we omit the temporal constraints, so we will work with P/A graphs only, but we will return to temporal constraints later in the paper. In particular, we are interested in finding whether it is possible to select a subset of nodes in such a way that they form a feasible graph according to the branching constraints. Formally, the selection of nodes can be described by an *assignment* of 0/1 values to nodes of a given P/A graph, where value 1 means that the node is selected and value 0 means that the node is not selected. The assignment is called *feasible* if

- in every parallel sub-graph all nodes are assigned the same value (both the principal node and all branching nodes are either all 0 or all 1),

- in every alternative sub-graph either all nodes (both the principal node and all branching nodes) are 0 or the principal node and exactly one branching node are 1 while all other branching nodes are 0.

Notice that the feasible assignment naturally describes one of the alternative processes in the P/A graph. For example, *weldRod* is present if and only if both *clearRod* and *assembleKit* are present (Figure 1). Similarly, *weldTube* is present if exactly one of nodes *buyTube* or *clearTube* is present (but not both). Though, the alternative branching is quite common in manufacturing scheduling, it cannot be described by binary logical constraints from MaScLib [12] or Extended Resource Constrained Project Scheduling Problem [9]. On the other hand, the branching constraints are specific logical relations that cannot capture all logical relations between the nodes.

Obviously, given an arbitrary P/A graph the assignment of value 0 to all nodes is always feasible. On the other hand, if some of the nodes are required to take value 1, then the existence of a feasible assignment is by no means obvious. Let us now formulate this decision problem formally.

Definition 3: Given a P/A graph G and a subset of nodes in G which are assigned to 1, the *P/A graph assignment problem* is "Is there a feasible assignment of 0/1 values to all nodes of G which extends the prescribed partial assignment?"

Intuition motivated by real-life examples says that it should not be complicated to select the nodes to form a valid process according to the branching constraints described above. The following proposition from [1] says the opposite.

Proposition 1: The P/A graph assignment problem is NP-complete.

Nevertheless, if we look back to the motivation example (Figure 1), we can see that the TNA has a specific topology which is, according to our experience, very typical for real-life processes. First, the process has usually one start point and one end point. Second, the graph is built by decomposing meta-processes into more specific processes until non-decomposable processes (operations) are obtained. There are basically three types of decomposition. The meta-process can split into two or more processes that run in a sequence, that is, after one process is finished, the subsequent process can start. The meta-process can split into two or more sub-processes that run in parallel, that is, all sub-processes start at the same time and the meta-process is finished when all sub-processes are finished. Finally, the meta-process may consists of several alternative sub-processes, that is, exactly one of these sub-processes is selected to do the job of the meta-process. Notice, that the last two decompositions have the same topology of the network, they only differ in the meaning of the branches in the network. Note finally, that we are focusing on modeling instances of processes with particular operations that will be allocated to time. Hence we do not assume loops that are sometimes used to model abstract processes. Figure 2 shows how the network from Figure 1 is constructed from a single arc by applying the above mentioned decomposition steps.

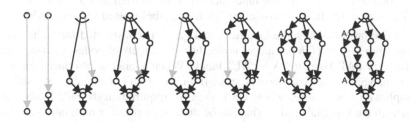

Fig. 2. Building a labeled nested graph

We will now formally describe this concept that we called nesting. The resulting network is called a Nested Temporal Network with Alternatives [2].

Definition 4: A directed graph $G = (\ \{s,e\},\ \{(s,e)\}\ )$ is a *(base) nested graph*. Let $G = (V, E)$ be a graph, $(x,y) \in E$ be one of its arcs, and $z_1,\ldots, z_k$ $(k > 0)$ be nodes such that no z_i is in V. If G is a nested graph (and $I = \{1,\ldots,k\}$) then graph $G' = (\ V \cup \{z_i \mid i \in I\},\ E \cup \{(x,z_i), (z_i,y) \mid i \in I\} - \{(x,y)\})$ is also a *nested graph*.

According to Definition 4, any nested graph can be obtained from the base graph with a single arc by repeated substitution of any arc (x,y) by a special sub-graph with k nodes (see Figure 3). Notice that a single decomposition rule covers both the serial process decomposition $(k = 1)$ and the parallel/alternative process decomposition $(k > 1)$. Though this definition is constructive rather than fully declarative, it is practically very useful. Namely, interactive process editors can be based on this definition so the users can construct only valid nested graphs by decomposing the base nested graph.

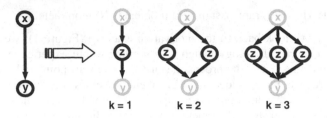

$$k = 1 \qquad k = 2 \qquad k = 3$$

Fig. 3. Arc decomposition in nested graphs

The directed nested graph defines topology of the nested P/A graph but we also need to annotate all fan-in and fan-out sub-graphs as either alternative or parallel sub-graphs. Moreover, we need to do the annotation carefully so the assignment problem can be solved easily for nested graphs and no node is inherently invalid. The idea is to annotate each node by input and output label which defines the type of branching (fan-in or fan-out sub-graph).

Definition 5: *Labeled nested graph* is a nested graph where each node has (possibly empty) input and output labels defined in the following way. Nodes s and e in the base nested graph and nodes z_i introduced during decomposition have empty initial labels. Let k be the number of nodes introduced when decomposing arc (x,y). If $k > 1$ then the output label of x and the input label of y are unified and set either to PAR or to ALT (if one of the labels is non-empty then this label is used for both nodes).

Figure 2 demonstrates how the labeled nested graph is constructed for the motivation example from Figure 1. In particular, notice how the labels of nodes are introduced (a semicircle for PAR label and A for ALT label). When a label is introduced for a node, it never changes in the generation process. If an arc (x, y) is being decomposed into a sub-graph with k new nodes where $k > 1$, then we require that the output label of x is unified with the input label of y. This can be done only if either both labels are identical or at least one of the labels is empty. It is easy to show that the second case always holds [2]. Now, we can formally introduce a nested P/A graph.

Definition 6: *A nested P/A graph* is obtained from a labeled nested graph by removing the labels and defining the fan-in and fan-out sub-graphs in the following way. If the input label of node x is non-empty then all arcs (y, x) form a fan-in sub-graph which is parallel for label PAR or alternative for label ALT. Similarly, nodes with a non-empty output label define fan-out sub-graphs. Each arc (x, y) such that both output label of x and input label of y are empty forms a parallel fan-in sub-graph.

Note, that requesting a single arc to form a parallel fan-in sub-graph is a bit artificial. We use this requirement to formally ensure that each arc is a part of some sub-graph which is required to show that a nested P/A graph is a P/A graph [2]. What is more interesting is that for Nested P/A Graphs the following proposition holds.

Proposition 2: The assignment problem for a nested P/A graph is tractable (can be solved in a polynomial time).

The formal proof in [2] is based on constructing a constraint model for nested P/A graphs where local (namely, arc) consistency, which is achievable in polynomial time, implies global consistency. If global consistency is achieved then the solution can be found using a backtrack-free depth-first search (provided that the problem is globally consistent, otherwise no solution exists). This constraint model is basically a (Berge acyclic) reformulation of the following straightforward model for the P/A graph assignment problem. Each node x is represented using a Boolean validity variable v_x, that is a variable with domain $\{0,1\}$. If the arc between nodes x and y is a part of some parallel sub-graph then we define the following constraint:

$$v_x = v_y. \tag{1}$$

If x is a principal node and $y_1,\ldots, y_k$ for some k are all branching nodes in some alternative sub-graph then the logical relation defining the alternative branching can be described using the following arithmetic constraint:

$$v_x = \Sigma_{j=1,\ldots,k} \, v_{y_j}. \tag{2}$$

Notice that if $k = 1$ then the constraints for parallel and alternative branching are identical (hence, it is not necessary to distinguish between them). Notice also that the arithmetic constraint for alternative branching together with the use of $\{0,1\}$ domains defines exactly the logical relation between the nodes – v_x is assigned to 1 if and only if exactly one of v_{y_j} is assigned to 1.

4 Temporal Constraints

So far, we focused merely on logical relations imposed by the branching constraints to show that logical reasoning is easy for nested P/A graphs (while it is hard for general P/A graphs). Now we return to the temporal constraints. Notice that the selected feasible set of nodes together with arcs between them forms a sub-graph of the original P/A graph. We require this sub-graph to be also *temporally feasible*, which means that all the temporal constraints between the valid nodes are satisfied in the sense of temporal networks [5]. Naturally, the logical and temporal reasoning is interconnected – if a temporal constraint between nodes x and y cannot be satisfied then (at least) one of the nodes must be invalid (it is assigned to 0). Before we go into technical details notice that if the temporal constraints are in the form of precedence relations or in general only minimal distances are specified in arcs (Figure 1) then temporal feasibility is trivially guaranteed thanks to acyclicity of TNAs (any node can be postponed in time). However, if deadlines are present (Figure 4) then temporal feasibility is not obvious similarly to situations when maximal distance between nodes is requested (for example, when cooling down restricts delays between operations).

Formally, we can extend the above logical constraint model by annotating each node i by temporal variable t_i indicating the position of the node in time. For simplicity reasons we assume that the domain of such variables is an interval $\langle 0, MaxTime \rangle$ of integers, where *MaxTime* is a large enough constant given by the user. Recall that the temporal relation between nodes i and j is described by a pair $[a_{i,j}, b_{i,j}]$. This relation can now be naturally represented using the following constraint:

$$v_i * v_j * (t_i + a_{i,j}) \le t_j \wedge v_i * v_j * (t_j - b_{i,j}) \le t_i. \tag{3}$$

If $b_{i,j} = \infty$ then the second part of the conjunction is omitted and similarly if $a_{i,j} = -\infty$ then the first part of conjunction is omitted. Notice that if any v_i or v_j equals zero (some involved node is invalid) then the constraint is trivially satisfied (we get $0 \leq t_j \wedge 0 \leq t_i$). If both v_i and v_j equal 1 then we get $(t_i + a_{i,j} \leq t_j \wedge t_j - b_{i,j} \leq t_i)$, which is exactly the simple temporal relation between nodes i and j. Figure 4 shows how the domains from the previous example (Figure 1) will look after filtering out the infeasible values by making the above constraint model arc consistent. We assume that *shipPiston* (the bottom node) is a valid node and *MaxTime* = 70. Black nodes are valid; validity of white nodes is not decided yet. Notice weak domain pruning of time variables in the white nodes caused by a disjunctive character of the problem. Actually, the left most path (with *buyTube*) cannot be selected due to time constraints but this is not discovered by making the constraints arc consistent.

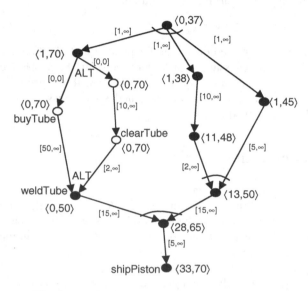

Fig. 4. Domain filtering using the constraint model

To improve domain filtering we propose to always propagate the temporal constraint even if the validity status of the node is not yet decided. If the temporal constraint is violated then we set some validity variable to 0 (if possible, otherwise a failure is detected). We will describe now the filtering rules that propagate changes of domains between the constrained variables, namely, the values that violate the constraint are removed from the domains. Let $d(x)$ be the domain of variable x, that is, a set of values, and for sets A and B, $A \bullet B = \{a \bullet b \mid a \in A \wedge b \in B\}$ for any binary operation $\bullet$ such as + or −.

Assume that arc (i, j) is a part of a parallel branching, so in the solution either both nodes i and j are valid and the temporal relation must hold, or both nodes are invalid and the temporal relation does not play any role (the domains of temporal variables are irrelevant provided that they are non-empty). Hence, we can always propagate the temporal relation provided that we properly handle its violation. Let $UP = d(t_j) \cap (d(t_i) + \langle a_{i,j}, b_{i,j} \rangle)$. The following filtering rule is applied whenever $d(t_i)$ changes:

$$d(t_j) \leftarrow UP \qquad \text{if } UP \neq \varnothing$$
$$d(v_j) \leftarrow d(v_j) \cap \{0\} \qquad \text{if } UP = \varnothing. \qquad (4)$$

Note that $UP = \varnothing$ means violation of the temporal relation which is accepted only if the nodes are invalid. If the nodes are valid then a failure is generated because the above rule makes the domain of the validity variable empty. Symmetrically, let $DOWN = d(t_i) \cap (d(t_j) - \langle a_{i,j}, b_{i,j} \rangle)$. The following filtering rule is applied whenever $d(t_j)$ changes:

$$d(t_i) \leftarrow DOWN \qquad \text{if } DOWN \neq \varnothing$$
$$d(v_i) \leftarrow d(v_i) \cap \{0\} \qquad \text{if } DOWN = \varnothing. \qquad (5)$$

The following example demonstrates the effect of above filtering rules. Assume that the initial domain of temporal variables is $\langle 0, 70 \rangle$, the validity of nodes is not yet decided, and there are arcs (i, j) and (j, k) with temporal constraints $[10, 30]$ and $[20, 20]$ respectively. The original constraints do not prune any domain, while our extended filtering rules set the domains of temporal variables t_i, t_j, and t_k to $\langle 0, 40 \rangle$, $\langle 10, 50 \rangle$, and $\langle 30, 70 \rangle$ respectively. If the initial domain is $\langle 0, 20 \rangle$ then the original constraints again prune nothing, while our extended filtering rules deduce that the participating nodes are invalid (we assume that logical constraints in the form $v_x = v_y$ are also present).

The propagation of temporal constraints in the alternative branching is more complicated because we do not know which arc is used in the solution. Therefore, the filtering rule uses a union of pruned domains proposed by individual arcs (from non-invalid nodes) which is similar to constructive disjunction of constraints. Let x be the principal node of a fan-in alternative sub-graph and $y_1,\dots, y_k$ be all branching nodes. We first show how domains of the branching nodes are propagated to the principal node. Let $UP = d(t_x) \cap \bigcup_{j = 1,\dots,k} \{(d(t_{y_j}) + \langle a_{y_j,x}, b_{y_j,x} \rangle) \mid d(v_{y_j}) \neq \{0\}\}$. The following filtering rule is applied whenever any $d(t_{y_j})$ or $d(v_{y_j})$ changes:

$$d(t_x) \leftarrow UP \qquad \text{if } UP \neq \varnothing$$
$$d(v_x) \leftarrow d(v_x) \cap \{0\} \qquad \text{if } UP = \varnothing. \qquad (6)$$

It may happen that set UP is not an interval but a set of intervals. Then we may use an interval hull which makes filtering less time and space consuming but a smaller number of inconsistent values is filtered out.

The propagation from $d(x)$ to $d(y_j)$ is done exactly like the $DOWN$ propagation described above (rule (5)) and similar filtering rules can be designed for fan-out alternative sub-graphs. Again, the main advantage of these rules is stronger pruning in comparison with the original constraints as we shall show using the example from Figure 4. In particular, if we propagate from $weldTube$ to $buyTube$ and $clearTube$, we obtain $\langle 0, 0 \rangle$ and $\langle 0, 48 \rangle$ as new domains of corresponding temporal variables. Now, if we propagate through the other alternative branching going to $buyTube$ from top, we deduce that this node is invalid because the corresponding temporal constraint is violated and hence $d(v_{buyTube}) \leftarrow \{0\}$. Consequently, all remaining nodes are valid and we achieved global consistency for both validity and temporal variables. Unfortunately, the proposed filtering rules do not guarantee global consistency in general. Figure 5 shows a nested TNA which is arc consistent, that is, the proposed filtering rules do not remove any inconsistent value from the current domains. However, there does not exist any solution to the problem.

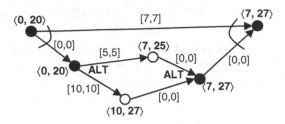

Fig. 5. Locally consistent nested TNA with no solution

As we shall show below, weak domain filtering of polynomial consistency techniques such as arc consistency is inevitable for (nested) TNAs because the problem of deciding existence of feasible assignment is in fact NP-complete.

Proposition 3: The problem of deciding whether there exists an assignment of times and 0/1 values to all nodes of the (nested) TNA in such a way that all temporal and branching constraints are satisfied is NP-complete.

Proof: The problem is obviously in NP, because it suffices to guess the assignment and test its feasibility, which can be done in linear time in the number of arcs. For the NP-hardness, we shall show that the subset sum problem, which is known to be NP-complete [6], can be reduced (in the polynomial time) to our assignment problem. The subset sum problem is this: given a set of positive integers Z_i and integer K, does the sum of some subset of $\{Z_i \mid i = 1,...,n\}$ equal to K? We can construct the following nested TNA, where the validity status of the black node is set to 1 and temporal annotation of arcs is [0,0] with the exception of n arcs annotated by $[Z_i, Z_i]$ and one arc annotated by [K,K] (Figure 6). Visibly, the subset sum problem has a solution if and only if there exists a feasible assignment of temporal and validity variables of the constructed nested TNA. The selection of the subset of integers is identical to the choice of alternative branches in the graph. The temporal constraints guarantee that the sum of selected integers equals K (the distance between the leftmost and rightmost node according to the top path). ∎

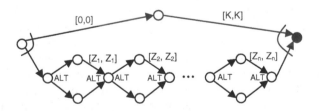

Fig. 6. Subset sum problem formulated as a Nested TNA

Nested TNAs can be seen from the point of view of disjunctive temporal networks [13] so a similar solving approach can be applied to obtain a consistent network. First, we find all solutions to the nested P/A assignment problem. Each solution defines a sub-graph of the Nested TNA which is a simple temporal network (STN) for which a consistency can be achieved in polynomial time [5] via path consistency or using all-pairs-shortest-path algorithms. So, in the second step we make all obtained STNs temporally consistent (if possible) or mark inconsistent STNs. Finally, we restrict the domains of temporal and validity variables in the following way. If a node is not present in any of consistent STNs then the node is made invalid. If a node is present in all consistent STNs then the node is made valid. Finally, the temporal domain for a non-invalid node is obtained by union of temporal domains of this node in all consistent STNs where the node is present. This constructive approach has been used in [8] for Temporal Planning Networks, but it has a problem if the number of generated STNs is too large. For example the problem from Figure 6 requires 2^n STNs to be explored. Hence, the worst case time of the method is exponential in the number of nodes.

We shall describe now a different algorithm which will compute the temporal domains of all vertices in such a way, that every value in every temporal domain is contained in some feasible solution. Let each edge (i, j) in a Nested TNA be labelled by set $S_{ij} \subseteq \langle 0, MaxTime \rangle$ of admissible values for the distance between nodes i and j. Initially, this set corresponds to interval $[a_{i,j}, b_{i,j}]$ specifying the temporal constraint. The proposed algorithm runs in two stages. In the first stage the sequence of decomposition steps used to construct the Nested TNA is followed in the reverse order (this sequence can be found algorithmically in polynomial time for any Nested TNA as shown in [2]). In each composition step in which a parallel or alternative sub-graph with principal vertices x and y and (not invalid) branching vertices $z_1, \ldots, z_k$ (if $k = 1$ then the type of branching is irrelevant) is replaced by a single edge (x, y), the set S_{xy} is computed in the following way:

- $S_{xy} = \cap_{i=1,\ldots,k} (S_{xz_i} + S_{z_iy})$ if the replaced sub-graph contains parallel branching,
- $S_{xy} = \cup_{i=1,\ldots,k} (S_{xz_i} + S_{z_iy})$ if the replaced sub-graph contains alternative branching.

We shall show later that the input TNA has a feasible solution if and only if the final base graph with only nodes s and e (into which the input TNA is composed in the end of the first stage) has a feasible solution. If there is no feasible solution, the algorithm terminates.

In the second stage of the algorithm, we compute restricted temporal constraints T_{ij} and restricted domains of temporal variables t_i containing only globally consistent values starting with the temporal domains in the base graph in the following way:

$$d(t_s) \leftarrow \langle 0, MaxTime \rangle \cap (\langle 0, MaxTime \rangle - S_{se})$$
$$d(t_e) \leftarrow \langle 0, MaxTime \rangle \cap (\langle 0, MaxTime \rangle + S_{se})$$
$$T_{se} \leftarrow S_{se}.$$

After that the base graph is decomposed again into the input graph. During each decomposition step in which a parallel or alternative sub-graph with principal vertices x and y and branching vertices $z_1, \ldots, z_k$ replaces a single edge (x, y), the sets T_{xz_i} and T_{z_iy} and the domains $d(z_i)$ for all $1 \leq i \leq k$ are computed in the following way:

- $T_{xz_i} = \{u \in S_{xz_i} \mid \exists v \in S_{z_iy} : u + v \in T_{xy}\}$
- $T_{z_iy} = \{v \in S_{z_iy} \mid \exists u \in S_{xz_i} : u + v \in T_{xy}\}$
- $d(t_{z_i}) = \{b \in \langle 0, MaxTime \rangle \mid \exists a \in d(x) \, \exists c \in d(y) : (b - a) \in T_{xz_i} \wedge (c - b) \in T_{z_iy} \}$

If $d(t_{z_i})$ is empty then vertex z_i is invalid so we can set $d(v_{z_i}) \leftarrow \{0\}$ and remove the vertex from the graph. This may happen only for alternative branching due to the way how S_{xy} is computed from S_{xz_i} and S_{z_iy}. Moreover, because S_{xy} is non-empty, at least one node z_i can still be valid so if any node is made invalid this is not propagated elsewhere in the graph. Notice also that $T_{xz_i} \subseteq S_{xz_i}$, $T_{z_iy} \subseteq S_{z_iy}$, and T_{xz_i} and T_{z_iy} contain only those pairs of values which sum up to some value in T_{xy}. We shall show now that only values participating in at least one feasible solution are ever inserted into the temporal domain of any vertex. First, let us define the notion of a feasible solution.

Definition 7: Let $G = (V, E)$ be a nested TNA where each edge $(i, j) \in E$ is labelled by set $S_{ij} \subseteq \langle 0, MaxTime \rangle$ of admissible values. An assignment $t : V \to \langle 0, MaxTime \rangle$ of temporal values (natural numbers) to vertices and $t : V \to \{0, 1\}\rangle$ of validity variables is called a feasible solution if for every edge $(i, j) \in E$ we have

$$(v_i * v_j = 1) \Rightarrow (t_j - t_i \in S_{ij}).$$

Remark: In the input TNA we assume that all sets S_{ij} are intervals, however for auxiliary TNA's constructed by the algorithm general sets will appear on newly introduced edges. Moreover, without lost of generality we can assume these sets to be within the interval $\langle 0, MaxTime \rangle$.

Lemma 1: Let $G = (V, E)$ be a TNA and let $G' = (V', E')$ be a TNA which originates from G by replacing a parallel/alternative sub-graph with principal vertices x and y and branching vertices $z_1, \ldots, z_k$ by edge (x, y). Then

a) if $t : V \to \langle 0, MaxTime \rangle$ is a feasible solution for G then $t' : V' \to \langle 0, MaxTime \rangle$ obtained by restricting t to V' (which is a subset of V) is a feasible solution for G', and

b) if $t' : V' \to \langle 0, MaxTime \rangle$ is a feasible solution for G' then there is a feasible solution $t : V \to \langle 0, MaxTime \rangle$ for G which is an extension of t' from V' to V.

Proof: (part a) The only edge that has to be checked is the newly introduced edge (x, y) originating from the composition operation. If $v_x = 0$ or $v_y = 0$ then the temporal constraint is satisfied trivially. So let us assume $v_x = v_y = 1$.

- If the replaced sub-graph contains parallel branching then $v_{z_i} = 1$ for all $1 \le i \le k$ and thus the feasibility of t implies $t_{z_i} - t_x \in S_{xz_i}$ and $t_y - t_{z_i} \in S_{z_iy}$ for all $1 \le i \le k$. Summing up these two relations gives us $t_y - t_x \in S_{xz_i} + S_{z_iy}$ (for every $1 \le i \le k$) and thus $t_y - t_x \in \cap (S_{xz_i} + S_{z_iy})$ which is by definition S_{xy} and so the temporal constraint on (x, y) is satisfied.

- If the replaced sub-graph contains alternative branching then $v_{z_i} = 1$ for exactly one index i, $1 \le i \le k$, and thus the feasibility of t implies $t_{z_i} - t_x \in S_{xz_i}$ and

$t_y - t_{z_i} \in S_{z_i y}$ for this particular i. Summing up these two relations gives us $t_y - t_x \in S_{xz_i} + S_{z_i y}$ (for the chosen i) and thus $t_y - t_x \in \cup (S_{xz_i} + S_{z_i y})$ which is by definition S_{xy} and so the temporal constraint on (x, y) is satisfied.

(part b) The edges that have to be checked in this case are the deleted edges (x, z_i) and (z_i, y). If $v_x = 0$ or $v_y = 0$ then also $v_{z_i} = 0$ for all $1 \leq i \leq k$ and all temporal constraints on the deleted edges are satisfied trivially. So let us again assume $v_x = v_y = 1$.

- If the replaced sub-graph contains parallel branching then $v_{z_i} = 1$ for all $1 \leq i \leq k$. The feasibility of t' implies $t'_y - t'_x \in \cap (S_{xz_i} + S_{z_i y})$. Let us pick an arbitrary index i, $1 \leq i \leq k$. The fact that $t'_y - t'_x \in S_{xz_i} + S_{z_i y}$ means that there exist $u \in S_{xz_i}$ and $v \in S_{z_i y}$ such that $t'_y - t'_x = u + v$. Now setting $t_{z_i} = t'_x + u = t'_y - v$ proves the desired result because $t_{z_i} - t'_x = u \in S_{xz_i}$ and $t_y - t_{z_i} = y \in S_{z_i y}$ and so the temporal constraints on (x, z_i) and (z_i, y) are satisfied.

- If the replaced sub-graph contains alternative branching then $v_{z_i} = 1$ for exactly one index i, $1 \leq i \leq k$, and we have a freedom of choice to determine which one (there must be at least one such vertex). The feasibility of t' implies $t'_y - t'_x \in \cup (S_{xz_i} + S_{z_i y})$. Let us fix an arbitrary index i, $1 \leq i \leq k$, such that $t'_y - t'_x \in S_{xz_i} + S_{z_i y}$ (at least one such i clearly exists) and set $v_{z_i} = 1$ and $v_{z_j} = 0$ for all $j \neq i$. This means that there exist $u \in S_{xz_i}$ and $v \in S_{z_i y}$ such that $t'_y - t'_x = u + v$. Now setting $t_{z_i} = t'_x + u = t'_y - v$ satisfies, in a similar fashion as above, the temporal constraints on (x, z_i) and (z_i, y). The temporal constraints on (x, z_j) and (z_j, y) for $j \neq i$ are satisfied trivially since $v_{z_j} = 0$. ∎

Corollary: The input TNA has a feasible solution if and only if the final base graph with only nodes s and e (into which the input TNA is composed in the end of the first stage) has a feasible solution.

Now let us state (and prove) the properties of the second stage of the algorithm.

Lemma 2: Let $G = (V, E)$ be a TNA and let $G' = (V', E')$ be a TNA which originates from G by replacing edge (x, y) by a parallel/alternative sub-graph with principal vertices x and y and branching vertices $z_1, \ldots, z_k$. Then if every pair of values $a \in d(t_x)$, $c \in d(t_y)$ such that $(c - a) \in T_{xy}$ participates in at least one feasible solution for G then also every pair of values $a \in d(t_x)$, $b \in d(t_{z_i})$ such that $(b - a) \in T_{xz_i}$ (and every pair of values $b \in d(t_{z_i})$, $c \in d(t_y)$ such that $(c - b) \in T_{z_i y}$) participates in at least one feasible solution for G'.

Proof: Let us consider an arbitrary $a \in d(t_x)$, $b \in d(t_{z_i})$ such that $(b - a) = u \in T_{xz_i}$. By the definition of T_{xz_i}, there exists (at least one) $v \in T_{z_i y}$ such that $(u + v) \in T_{xy}$. Let us define $W = \{v \in T_{z_i y} \mid u + v \subset T_{xy}\}$ and $C = \{a + u + v \mid v \in W\}$. By the definition of $d(t_{z_i})$ there must be at least one $c \in C$ such that $c \in d(t_y)$. However, now $(c - a) = (u + v) \in T_{xy}$ and so, by the assumption, the pair of values a, c participates in at least one feasible solution for G. Clearly, this solution can be extended by value $b \in d(t_{z_i})$ without violating the temporal constraints on edges (x, z_i) and (z_i, y). ∎

Proposition 4: After the second stage of the algorithm terminates, the temporal domains in the input graph fulfil global consistency.

Proof: The base graph clearly satisfies the assumptions of Lemma 2, namely every pair of values $x \in d(t_s)$ and $y \in d(t_e)$ such that $(y - x) \in T_{se}$ participate in at least one feasible solution for the base graph. Thus, due to Lemma 2, also every graph obtained by a single decomposition step satisfies the statement of Lemma 2. Moreover, every value b newly introduced into the domain $d(t_{zi})$ has at least one value a in $d(t_x)$ and one value c in $d(t_y)$ such that $(b - a) \in T_{xzi}$ and $(c - b) \in T_{ziy}$. Thus every such value b participates in at least one feasible solution. ∎

Since we give no implementation details here, it is not possible to determine the exact time complexity of the presented algorithm. However, it should be clear, that any reasonable implementation will work in time polynomial in the size of the input TNA and the upper bound *MaxTime*, thus providing a pseudo-polynomial algorithm with respect to the size of input data (the constant *MaxTime* is part of the input but coded in binary and thus taking *log MaxTime* bits).

5 Conclusions

The paper studies temporal reasoning in Temporal Networks with Alternatives which are useful to model alternative process in production scheduling. We showed that adding simple temporal constraints to Nested TNAs makes the problem of deciding existence of logically and temporally feasible solution NP-complete. We presented a straightforward constraint model and stronger filtering rules that can remove, via arc consistency, some infeasible values from variables' domains, but still cannot guarantee global consistency. We also presented an algorithm for achieving global consistency with pseudo-polynomial time complexity. Note that this algorithm is applicable only to Nested TNAs while the proposed filtering rules work for any TNA. This paper focuses on theoretical aspects of reasoning with Nested TNAs, the next step is empirical evaluation of the presented techniques.

Acknowledgments. The research is supported by the Czech Science Foundation under the contract no. 201/07/0205. We would like to thank anonymous reviewers for their valuable comments.

References

1. Barták, R., Čepek, O.: Temporal Networks with Alternatives: Complexity and Model. In: Proceedings of the Twentieth International Florida AI Research Society Conference (FLAIRS), pp. 641–646. AAAI Press, Menlo Park (2007)
2. Barták, R., Čepek, O.: Nested Temporal Networks with Alternatives, Papers from the 2007 AAAI Workshop on Spatial and Temporal Reasoning, Technical Report WS-07-12, pp. 1–8. AAAI Press, Menlo Park (2007)
3. Beck, J.C., Fox, M.S.: Scheduling Alternative Activities. In: Proceedings of AAAI 1999, pp. 680–687. AAAI Press, Menlo Park (1999)

4. Blythe, J.: An Overview of Planning Under Uncertainty. AI Magazine 20(2), 37–54 (1999)
5. Dechter, R., Meiri, I., Pearl, J.: Temporal Constraint Networks. Artificial Intelligence 49, 61–95 (1991)
6. Garey, M.R., Johnson, D.S.: Computers and Intractability: A Guide to the Theory of NP-Completeness. W. H. Freeman and Company, San Francisco (1979)
7. Hamadi, Y.: Cycle-cut decomposition and log-based reconciliation. In: ICAPS Workshop on Connecting Planning Theory with Practice, pp. 30–35 (2004)
8. Kim, P., Williams, B., Abramson, M.: Executing Reactive, Model-based Programs through Graph-based Temporal Planning. In: Proceedings of International Joint Conference on Artificial Intelligence (IJCAI) (2001)
9. Kuster, J., Jannach, D., Friedrich, G.: Handling Alternative Activities in Resource-Constrained Project Scheduling Problems. In: Proceedings of Twentieth International Joint Conference on Artificial Intelligence (IJCAI 2007), pp. 1960–1965 (2007)
10. Laborie, P.: Resource temporal networks: Definition and complexity. In: Proceedings of the 18th International Joint Conference on Artificial Intelligence, pp. 948–953 (2003)
11. Moffitt, M.D., Peintner, B., Pollack, M.E.: Augmenting Disjunctive Temporal Problems with Finite-Domain Constraints. In: Proceedings of the 20th National Conference on Artificial Intelligence (AAAI 2005), pp. 1187–1192. AAAI Press, Menlo Park (2005)
12. Nuijten, W., Bousonville, T., Focacci, F., Godard, D., Le Pape, C.: MaScLib: Problem description and test bed design (2003), http://www2.ilog.com/masclib
13. Stergiou, K., Koubarakis, M.: Backtracking algorithms for disjunctions of temporal constraints. In: Proceedings of the 15th National Conference on Artificial Intelligence (AAAI 1998), pp. 248–253. AAAI Press, Menlo Park (1998)
14. Tsamardinos, Vidal, T., Pollack, M.E.: CTP: A New Constraint-Based Formalism for Conditional Temporal Planning. Constraints 8(4), 365–388 (2003)

SCLP for Trust Propagation in Small-World Networks*

Stefano Bistarelli[1,2] and Francesco Santini[2,3]

[1] Dipartimento di Scienze, Università "G. d'Annunzio" di Chieti-Pescara, Italy
bista@sci.unich.it
[2] Istituto di Informatica e Telematica (CNR), Pisa, Italy
{stefano.bistarelli,francesco.santini}@iit.cnr.it
[3] IMT - Institute for Advanced Studies, Lucca, Italy
f.santini@imtlucca.it

Abstract. We propose *Soft Constraint Logic Programming* based on semirings as a mean to easily represent and evaluate trust propagation in small-world networks. To attain this, we model the trust network adapting it to a weighted *and-or* graph, where the weight on a connector corresponds to the trust and confidence feedback values among the connected nodes. Semirings are the parametric and flexible structures used to appropriately represent trust metrics. Social (and not only) networks present small-world properties: most nodes can be reached from every other by a small number of hops. These features can be exploited to reduce the computational complexity of the model. In the same model we also introduce the concept of *multitrust*, which is aimed at computing trust by collectively involving a group of trustees at the same time.

1 Introduction

Decentralized trust management [16] provides a different paradigm of security in open and widely distributed systems where it is not possible to rely solely on traditional security measures as cryptography. The reasons usually are that the nodes appear and disappear from the community, span multiple administrative domains, their direct interactions are limited to a small subset of the total number of nodes and, moreover, there is no globally trusted third party that can supervise the relationships. For this reason an expressive computational model is needed to derive a trust value among the individuals of a community, represented as a trust network, in the following abbreviated as *TN*.

Three main contributions are given in this paper: first of all we propose the concept of *multitrust* [8], i.e. when the relationship of trust concerns one trustor and multiple trustees in a correlated way (the name recalls the *multicast* delivery scheme in networks). An example in peer-to-peer networks is when we download a file from multiple sources at the same time, and we need a reliability feedback for the whole download process. This result depends on the integrated characteristics of all the sources.

---

* Supported by the MIUR PRIN 2005-015491.

F. Fages, F. Rossi, and S. Soliman (Eds.): CSCLP 2007, LNAI 5129, pp. 32–46, 2008.

Secondly, we outline a model to solve trust propagation: we represent TNs (the same model applies also to related terms in literature as trust graph, web of trust or social network [25]) as *and-or* graphs [17] (i.e. hypergraphs), mapping individuals to nodes and their relationships to directed connectors. The *and* connectors (i.e. hyperarcs) represent the event of simultaneously trusting a group of individuals at the same time. The costs of the connectors symbolize how trustworthy the source estimates the destination nodes, that is a *trust value*, and how accurate is this trust opinion, i.e. a *confidence value*. Afterwards, we propose the *Soft Constraint Logic Programming* (SCLP) framework [1,5] as a convenient declarative programming environment in which to solve the trust propagation problem for multitrust. In SCLP programs, logic programming is used in conjunction with soft constraints, that is, constraints which have a preference level associated to them. In particular, we show how to translate the *and-or* graph obtained in the first step into a SCLP program, and how the semantics of such a program computes the best trust propagation tree in the corresponding weighted *and-or* graph. SCLP is based on the general structure of a *c-semiring* [1] (or simply, semiring) with two operations $\times$ and $+$. The $\times$ is used to combine the preferences, while the partial order defined by $+$ (see Section 2) is instead used to compare them.

Therefore, we can take advantage of the semiring structure to model and compose different trust metrics. SCLP is also parametric w.r.t. the chosen semiring: the same program deals with different metrics by only choosing the proper semiring structure. In [7], a similar model has been proposed for routing.

We practically solve the problem with *CIAO Prolog* [9] (modelling SCLP), a system that offers a complete Prolog system supporting ISO-Prolog , but, at the same time its modular design allows both restricting and extending the basic language. Thus, it allows both to work with subsets of Prolog and to work with programming extensions implementing functions, higher-order (with predicate abstractions), constraints, fuzzy sets, objects, concurrency, parallel and distributed computations, sockets, interfaces to other programming languages (C, Java, Tcl/Tk) and relational databases and many more.

The third and final contribution is represented by a practical implementation of the framework on a random small-world network [14] generated with the *Java Universal Network/Graph Framework* (JUNG) [18]. The small-world phenomenon describes the tendency for each entity in a large system to be separated from any other entity by only a few hops. Moreover, these networks a high clustering coefficient, which quantifies how close a vertex and its neighbors are from being a clique (i.e. a high coefficient suggests a clique). As a result, the problem can be divided in subproblems, each of them representing the topology of a clique, and then trying to connect these group of nodes together. The small number of hops allows to cut the solution search after a small threshold, thus improving the search even in wide networks.

This paper is organized as follows: In Sec. 2 we present some background information about trust metrics, the small-world phenomenon in social networks and the SCLP framework. Section 3 depicts how to represent a TN with an *and-or* graph, while in Sec. 4 we describe the way to pass from *and-or* graphs to SCLP programs, showing that the semantic of SCLP program is able to compute the best trust propagation in the corresponding *and-or* graph. In Sec. 5 we describe the practical implementation of the framework for a small-world network, and we suggest how to improve the performance. At last, Section 6 draws the final conclusions and outlines intentions about future works.

2   Background

Trust, Multitrust and Metrics. No universal agreement on the definition of trust and reputation concepts has been yet reached in the trust community [16]. However, we adopt the following definitions: trust describes a nodes belief in another nodes capabilities, honesty and reliability based on its own direct experiences, while reputation is based on recommendations received also from other nodes. Even if closely related, the main difference between trust and reputation is that trust systems produce a score that reflects the relying party's subjective view of an entity's trustworthiness, whereas reputation systems produce an entity's (public) reputation score as seen by the whole community.

Trust and reputation ranking metrics have primarily been used for public key certification, rating and reputation systems part of online communities, peer-to-peer networks, semantic web and also mobile computing fields [16,22,25]. Each of these scenarios favors different trust metrics. Trust metrics are used to predict trust scores of users by exploiting the transitiveness property of relationships (thus, we are considering transitive trust chains): if two nodes, say node A and node C in Fig. 1a, do not have a direct edge connecting them, the TN can be used to generate an inferred trust rating. A TN represents all the direct trust relationship in a community. An example of a classical TN is provided in Fig. 1a, where we can see that trust is usually represented as a 1-to-1 relationship between only two individuals: the edges are directed from the trustor to the trustee. If node A knows node B, and node B knows node C, then A can use the path to compose the inferred rating for C: therefore, we use transitive relationships. This process is called *trust propagation* by concatenation, and it is a necessary requirement since in most settings a user has a direct opinion only about a very small portion of nodes in the TN. Therefore, trust needs to be granted also by basing on third-party recommendations: if A trusts B, she/he can use the recommendation about C provided by B [16]. How to compose this information depends on the trust metrics of the links, i.e. it specifically depends on the problem [16] (e.g. by multiplying together the trust scores of the links A-B and B-C).

We introduce the concept of multitrust [8], which extends the usual trust relationship from couples of individuals to one trustor and multiple trustees in a correlated way: is the set of entities is denoted with E, the multitrust relationship R_{mt} involves a trustor $t \in E$ and a set of trustees $T \subset E$. The correlation in R_{mt} can be defined in terms of time (e.g. at the same time), modalities (e.g. with the same behavior) or collaboration among the trustees in T w.r.t. t. For example if we consider time, the trustor could simultaneously trust multiple trustees, or, considering instead a modality example, the trustor could contact the trustees with the same communication device, e.g. by phone. Consequently, this trust relation R_{mt} is 1-to-n (no more 1-to-1 as in all the classical trust systems [25]) and can be created by concurrently involving all the interested parties in a shared purpose. A general application can be for *team effectiveness* [10]: suppose we have a decentralized community of open-source programmers and we want to know if a subset of them can be reliably assigned to a new project.

A team of 3 programmers, for example, could significantly enhance the software product since we suppose they will accurately collaborate together by joining their skills and obtaining a better result w.r.t. 3 independent developers. Thus, the group will be more trustworthy than the single individuals, and even the final trustees will benefit from this group collaboration: they will be reached with an higher score during the propagation of trust in the TN.

Small-World Networks and Trust. A social network, where nodes represent individuals and edges represent their relationships, exhibits the small-world phenomenon if any two individuals in the network are likely to be connected through a short sequence of intermediate acquaintances. In [24] the authors observe that such graphs have a high clustering coefficient (like regular graphs) and short paths between the nodes (like random graphs).

These networks are divided in sub-communities (i.e. in clusters) where few individuals, called the *pivots* [13], represent the bridges towards different groups. These connections are termed *weak ties* in the sociology literature [13], as opposed to *strong ties* that connect a vertex to others in its own sub-community. Weak ties are important because the individuals inside other communities will bring in greater value due to different knowledge and perspectives, while people in the same group would generally tend to have the same knowledge. An example of small-world network is represented in Sec. 5.

Soft Constraint Logic Programming. The SCLP framework [1,5,12], is based on the notion of *c-semiring* introduced in [4,6]. A semiring S is a tuple $\langle A, +, \times, 0, 1 \rangle$ where A is a set with two special elements $(0, 1 \in A)$ and with two operations $+$ and $\times$ that satisfy certain properties: $+$ is defined over (possibly infinite) sets of elements of A and thus is commutative, associative, idempotent, it is closed and 0 is its unit element and 1 is its absorbing element; $\times$ is closed, associative, commutative, distributes over $+$, 1 is its unit element, and 0 is its absorbing element (for the exhaustive definition, please refer to [6]). The $+$ operation defines a partial order $\leq_S$ over A such that $a \leq_S b$ iff $a + b = b$; we say that $a \leq_S b$ if b represents a value *better* than a. Other properties related to the two operations are that $+$ and $\times$ are monotone on $\leq_S$, 0 is its minimum and 1 its maximum, $\langle A, \leq_S \rangle$ is a complete lattice and $+$ is its lub. Finally, if $\times$ is idempotent, then $+$ distributes over $\times$, $\langle A, \leq_S \rangle$ is a complete distributive lattice and $\times$ its glb.

Semiring-based Constraint Satisfaction Problems (SCSPs) [1] are constraint problems where each variable instantiation is associated to an element of a c-semiring A (to be interpreted as a cost, level of preference, ...), and constraints are combined via the $\times$ operation and compared via the $\leq_S$ ordering. Varying the set A and the meaning of the $+$ and $\times$ operations, we can represent many different kinds of problems, having features like fuzziness, probability, and optimization. Moreover, since the cartesian product of semirings is still a semiring [1], trust can be propagated by considering several criteria (i.e. metrics) at the same time, trying to optimize all the different scores.

Constraint Logic Programming (CLP) [15] extends Logic Programming by replacing term equalities with constraints and unification with constraint solving. The SCLP framework extends the classical CLP formalism in order to be able to handle also SCSP [4,6] problems. In passing from CLP to SCLP languages, we replace classical constraints with the more general SCSP constraints where we are able to assign a *level of preference* to each instantiated constraint (i.e. a ground atom). To do this, we also modify the notions of interpretation, model, model intersection, and others, since we have to take into account the semiring operations and not the usual CLP operations. The fact that we have to combine several refutation paths (a refutation is a finite derivation and the corresponding semiring value a [1]) when we have a partial order among the elements of the semiring (instead of a total one), can be fruitfully used in the context of this paper when we have an graph/hypergraph problems with incomparable costs associated to the edges/connectors. In fact, in the case of a partial order, the solution of the problem of finding the best path/tree should consist of all those paths/trees whose cost is not "dominated" by others.

Table 1. A simple example of an SCLP program

```
s(X)    :- p(X,Y).
p(a,b)  :- q(a).
p(a,c)  :- r(a).
q(a)    :- t(a).
t(a)    :- 2.
r(a)    :- 3.
```

A simple example of a SCLP program over the semiring $\langle N, min, +, +\infty, 0\rangle$, where N is the set of non-negative integers and $D = \{a, b, c\}$, is represented in Tab. 1. The intuitive meaning of a semiring value like 3 associated to the atom $r(a)$ (in Tab. 1) is that $r(a)$ costs 3 units. Thus the set N contains all possible costs, and the choice of the two operations min and $+$ implies that we intend to minimize the sum of the costs. This gives us the possibility to select the atom instantiation which gives the minimum cost overall. Given a goal like $s(x)$ to this program, the operational semantics collects both a substitution for x (in this case, $x = a$) and also a semiring value (in this case, 2) which represents the minimum cost among the costs for all derivations for $s(x)$. To find one of these solutions, it starts from the goal and uses the clauses as usual in logic programming, except that at each step two items are accumulated and combined with the current state: a substitution and a semiring value (both provided by the used clause). The combination of these two items with what is contained in the current goal is done via the usual combination of substitutions (for the substitution part) and via the multiplicative operation of the semiring (for the semiring value part), which in this example is the arithmetic $+$. Thus, in the example of goal $s(X)$, we get two possible solutions, both with substitution $X = a$ but with two different semiring values: 2 and 3. Then, the combination of such two solutions via the min operation give us the semiring value 2.

3   From Trust Networks to *and-or* Graph

An *and-or* graph [17] is defined as a special type of hypergraph. Namely, instead of arcs connecting pairs of nodes there are hyperarcs connecting an n-tuple of nodes ($n = 1, 2, 3, \dots$). The arcs are called *connectors* and they must be considered as directed from their first node to all the other nodes in the n-tuple. Formally an *and-or* graph is a pair $G = (N, C)$, where N is a set of *nodes* and C is a set of connectors defined as $C \subseteq N \times \bigcup_{i=0}^{k} N^i$.

When $k > 1$ we have an *and* connector since it reaches multiple destinations at the same time; all the different connectors rooted in the same n_i node can be singly chosen, i.e. *or* connectors. Note that the definition allows 0-connectors, i.e. connectors with one input and no output node. In the following of the explanation we will also use the concept of *and* tree [17]: given an *and-or* graph G, an *and* tree H is a *solution tree of G with start node n_r*, if there is a function g mapping nodes of H into nodes of G such that: *i)* the root of H is mapped in n_r, and *ii)* if $(n_{i_0}, n_{i_1}, \dots, n_{i_k})$ is a connector of H, then $(g(n_{i_0}), g(n_{i_1}), \dots, \dots, g(n_{i_k}))$ is a connector of G.

Informally, a solution tree of an *and-or* graph is analogous to a path of an ordinary graph: it can be obtained by selecting exactly one outgoing connector for each node. If all the chosen connectors are 1-connectors, then we obtain a plain path and not a tree.

In Fig. 1b we directly represent a TN for multitrust as a weighted *and-or* graph, since for its characteristics, this translation is immediate. Each of the individuals can be easily cast in a corresponding node of the *and-or* graph. In Fig. 1b we represent our trustor as a black node (i.e. n_1) and the target trustees as two concentric circles (i.e. n_4 and n_5). Nodes n_2 and n_3 can be used to propagate trust.

To model the trust relationship between two nodes we use 1-connectors, which correspond to usual TN arcs: the 1-connectors in Fig. 1b are (n_1, n_2), (n_1, n_3), (n_2, n_3), (n_2, n_4), (n_3, n_4), (n_3, n_5), (n_4, n_5). We remind that the connectors are directed, and thus, for example the connector (n_4, n_5) means that the input node n_4 trusts the individual represented by n_5. Moreover, since we are now dealing with multitrust, we need to represent the event of trusting more individuals at the same time. To attain this, in Fig. 1b we can see the three 2-connectors (n_1, n_2, n_3), (n_2, n_3, n_4) and (n_3, n_4, n_5): for example, the first of these hyperconnectors defines the possibility for n_1 to trust both n_2 and n_3 in a correlated way. In Fig. 1b we draw these n-connectors (with $n > 1$) as curved oriented arcs where the set of their output nodes corresponds to the output nodes of the 1-connectors traversed by the curved arc. Considering the ordering of the nodes

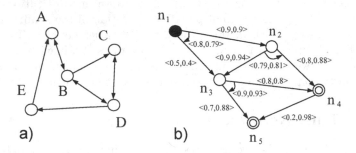

Fig. 1. *a)* A classical trust network, and *b)* an *and-or* graph representing multitrust: the weights on the connectors represent trust and confidence values (i.e. $\langle t, c \rangle$)

in the tuple describing the connector, the input node is at the first position and the output nodes (when more than one) follow the orientation of the related arc in the graph (in Figure 1b this orientation is lexicographic). Notice that in the example we decided to use connectors with dimension at most equal to 2 (i.e. 2-connectors) for sake of simplicity. However it is possible to represent whatever cardinality of trust relationship, that is among a trustor and n trustees (i.e. with a n-connector).

So far, we are able to represent an entire TN with a weighted *and-or* graph, but still we need some algebraic framework to model our preferences for the connectors, to use during trust propagation as explained in the following. For this purpose we decided to use the semiring structure. Each of the connectors in Fig. 1b is labeled with a couple of values $\langle t, c \rangle$: the first component represents a trust value in the range $[0, 1]$, while the second component represents the accuracy of the trust value assignment (i.e. a *confidence* value), and it is still in the range $[0, 1]$. This parameter can be assumed as a *quality* of the opinion represented instead by the trust value; for example, a high confidence could mean that the trustor has interacted with the target for a long time and then the correlated trust value is estimated with precision. A trust value close to 1 indicates that the output nodes of the connector have gained a good feedback in terms of their past performance and thus are more trustworthy, whereas a low trust value means the nodes showed relatively poor QoS in the past and are rated with low score. In general, we could have trust expressed with a k-dimensional vector representing k different metrics; in this example we have 2-dimensional vectors. Therefore, the semiring we use to propagate trust in the network is based on the *path semiring* [21]: $S_{trust} = \langle \langle [0,1], [0,1] \rangle, +_p, \times_p, \langle 0,0 \rangle, \langle 1,1 \rangle \rangle$, where

$$\langle t_i, c_i \rangle +_p \langle t_j, c_j \rangle = \begin{cases} \langle t_i, c_i \rangle & \text{if } c_i > c_j, \\ \langle t_j, c_j \rangle & \text{if } c_i < c_j, \\ \langle max(t_i, t_j), c_i \rangle & \text{if } c_i = c_j. \end{cases}$$

$$\langle t_i, c_i \rangle \times_p \langle t_j, c_j \rangle = \langle t_i t_j, c_i c_j \rangle$$

Along the same path, the $\times_p$ computes the scalar product of both trust and confidence values, and since the considered interval is $[0, 1]$, they both decrease when aggregated along a path. When paths are instead compared, $+_p$ chooses the one with the highest confidence. If the two opinions have equal confidences but different trust values, $+_p$ picks the one with the highest trust value. In this way, the precision of the information is more important than the information itself. If the k-dimensional costs of the connectors are not elements of a totally ordered set (therefore, not in our trust/confidence example), it may be possible to obtain several Pareto-optimal solutions.

Notice that other semirings can be used to model other trust metrics: for example, the *Fuzzy Semiring* $\langle [0, 1], \max, \min, 0, 1 \rangle$ can be used if we decide that the score of a trust chain corresponds to the weakest of its links. Or we can select the *Weighted Semiring*, i.e. $\langle \mathcal{R}^+, \min, +, \infty, 0 \rangle$, to count negative referrals in reputation systems as in e-Bay [16].

Collecting the trust values to assign to the labels of the connectors is out of the scope of this work, but they can be described in terms of specificity/generality dimensions (if we relay on one or more aspects) and subjective/objective dimensions (respectively personal, as e-Bay, or formal criteria, as credit rating) [16]. However, for n-connectors with $n \geq 2$, we can suppose also the use of a composition operation $\circ$ which takes n k-dimensional trust metric vectors (e.g. $tvalue_1, \ldots, tvalue_n$) as operands and returns the estimated trust value for the considered n-connector $(tvalue_{nc})$: $\circ$ $(tvalue_1, tvalue_2, \ldots, tvalue_n) \rightarrow tvalue_{nc}$.

This $\circ$ operation can be easily found for objective ratings, since they are the result of applying formal aspects that have been clearly defined, while automating the computation of subjective ratings is undoubtedly more difficult. Notice also, as said before, that such a $\circ$ operation is not only a plain "addition" of the single trust values, but it must take into account also the "added value" (or "subtracted value") derived from the combination effect. For example, considering the connector (n_3, n_4, n_5) in Fig. 1b, its cost, i.e. $\langle 0.9, 0.93 \rangle$, significantly benefits from simultaneously trusting n_4 and n_5, since both the trust/confidence values of (n_3, n_4) and (n_3, n_5) are sensibly lower (i.e. respectively $\langle 0.8, 0.8 \rangle$ and $\langle 0.7, 0.88 \rangle$). The reason could be that n_3 has observed many times the collaboration between n_4 and n_5 (i.e. a high confidence value) and this collaboration is fruitful. On the other hand, n_2 does not consider n_3 and n_4 to be so "collaborative" since the trust label of (n_2, n_3, n_4), i.e. $\langle 0.8, 0.81 \rangle$, is worse than the costs of (n_2, n_3) and (n_2, n_4) (i.e. $\langle 0.9, 0.94 \rangle$ and $\langle 0.8, 0.88 \rangle$). In the example in Fig. 1b we supposed to use subjective ratings, and therefore the trust values for 2-connectors do not follow any specific $\circ$ function. An example of objective rating could be the average mean function for both trust and confidence values of all the composing 1-connectors.

Notice that sometimes trust is computed by considering all the paths between two individuals and then by applying a function in order to find a single result [22] (e.g. the mean of the trust scores for all the paths). This could be accomplished by using the *expectation* semiring [11], where the + operation of the semiring is used to aggregate the trust values across paths, as proposed in [21]. In this paper, we decide to keep + as a "preference" operator for distinct paths (as proposed for classical SCLP, see Sec. 2) in order to choose the best one, since in Sec. 5.1 we suggest how to reduce the complexity of the framework by visiting less paths as possible. Thus, aggregating the trust values of the paths is not so meaningful when trying to reduce the number of visited paths at the same time.

4   *and-or* Graphs Using SCLP

In this Section, we explain how to represent *and-or* graphs with a program in SCLP. This decision is derived from two important features of this programming framework: *i)* SCLP is a declarative programming environment and, thus, is relatively easy to specify a lot of different problems; *ii)* the c-semiring structure is a very parametric tool where to represent several and different trust metrics. As a translation example we consider the *and-or* graph in Fig. 1b: by only changing the facts in the program, it is possible to translate every other tree.

Using this framework, we can easily find the best trust propagation over the hypergraph built in Sec. 3. In fact, our aim is to find the best path/tree simultaneously reaching all the desired trustees, which is only one of the possible choices when computing trust [22]: according to *multipath propagation*, when multiple propagation paths (in this case, trees) exist between A and C (in this case, several trustees at the same time), all their relative trust scores can be composed together in order to have a single result balanced with every opportunity. To attain multipath propagation we need to use the *expectation* semiring [11] as explained in Sec. 3.

In SCLP a clause like $c(n_i, [n_j, n_k])$:- *tvalue*, means that the graph has connector from n_i to nodes n_j and n_k with *tvalue* cost. Then, other SCLP clauses can describe the structure of the path/tree we desire to search over the graph. Notice that possible cycles in the graph are automatically avoided by SCLP, since the $\times$ of the semiring is a monotonic operation.

As introduced in Sec. 1, we use CIAO Prolog [9] as the system to practically solve the problem. CIAO Prolog has also a fuzzy extension, but it does not completely conform to the semantic of SCLP defined in [5] (due to interpolation in the interval of the fuzzy set). For this reason, we inserted the cost of the connector in the head of the clauses, differently from SCLP clauses which have the cost in the body of the clause.

From the *and-or* graph in Fig. 1b we can build the corresponding CIAO program of Tab. 2 as follows. First, we describe the connectors of the graph with facts like

$$connector(trustor, [trustees\_list], [trust\_value, condifence\_value])$$

e.g. the fact $connector(n_1, [n_2, n_3], [0.8, 0.79])$ represents the connector of the graph (n_1, n_2, n_3) with a trust/confidence value of $\langle 0.8, 0.79 \rangle$ (n_i represents the name of the node). The set of connector facts is highlighted as *Connectors* in Tab. 2, and represents all the trust relationships of the community. The *Leaves* facts of Tab. 2 represent the terminations for the Prolog rules. Their cost must not influence the final trust score, and then it is equal to the unit element

Table 2. The CIAO program representing the *and-or* graph in Fig. 1b

```
:- module(trust,_,_).
:- use_module(library(lists)).
:- use_module(library(aggregates)).
:- use_module(library(sort)).
```

times
```
times([T1, C1], [T2, C2], [T, C]) :-
    T is (T1 * T2),
    C is (C1 * C2).
```

plus
```
plus([], MaxSoFar, MaxSoFar).

plus([[T,C]|Rest], [MT,MC], Max):-
    C > MC, plus(Rest, [T,C], Max).

plus([[T,C]|Rest], [MT,MC], Max):-
    C = MC, T > MT,
    plus(Rest, [T,C], Max).

plus([[T,C]|Rest], [MT,MC], Max):-
    C < MC,
    plus(Rest, [MT,MC], Max).

plus([[T,C]|Rest], [MT,MC], Max):-
    C = MC,
    T < MT,
    plus(Rest, [MT,MC], Max).
```

trust
```
trust(X, Y, Max):-
    findall([T,C], trustrel(X, Y, [T,C]), L1),
    plus(L1,[0,0],Max).
```

Leaves
```
leaf([n1], [1,1]).
leaf([n2], [1,1]).
leaf([n3], [1,1]).
leaf([n4], [1,1]).
leaf([n5], [1,1]).
```

Connectors
```
connector(n1,[n2], [0.9,0.9]).
connector(n1,[n3], [0.5,0.4]).
connector(n1,[n2,n3], [0.8,0.79]).
connector(n2,[n3], [0.9,0.94]).
connector(n2,[n4], [0.8,0.88]).
connector(n2,[n3,n4], [0.8,0.82]).
connector(n3,[n4], [0.8,0.8]).
connector(n3,[n5], [0.7,0.88]).
connector(n3,[n4,n5], [0.9,0.93]).
connector(n4,[n5], [0.2,0.98]).
```

1)
```
trustrel(X,[X], [T,C]):-
    leaf([X], [T,C]).
```

2)
```
trustrel(X, Z, [T,C]):-
    connector(X,W, [T1,C1]),
    trustrelList(W, Z, [T2,C2]),
    times([T1,C1], [T2,C2], [T,C]).
```

3) `trustrelList([],[], [1,1]).`

4)
```
trustrelList([X|Xs],Z, [T,C]):-
    trustrel(X, Z1, [T1,C1]),
    append(Z1, Z2, Z),
    trustrelList(Xs, Z2, [T2,C2]),
    times([T1,C1], [T2,C2], [T,C]).
```

of the $\times$ operator of the S_{trust} semiring presented in Sec. 3, i.e. $\langle 1, 1 \rangle$. The *times* and *plus* clauses in Tab. 2 respectively mimic the $\times$ and $+$ operation of $S_{trust} = \langle \langle [0,1], [0,1] \rangle, +_p, \times_p, \langle 0,0 \rangle, \langle 1,1 \rangle \rangle$ explained in Sec. 3. The *trust* clause is used as the query to compute trust in the network: it collects all the results for the given source and destinations, and then finds the best trust/confidence couple by using the *plus* clauses.

At last, the rules 1-2-3-4 in Tab. 2 describe the structure of the relationships we want to find over the social network: with these rules it is possible to found both 1-to-1 relationships (i.e. for classical trust propagation) or 1-to-n relationships (i.e. for multitrust propagation, described in Sec. 2). *Rule 1* represents a relationship made of only one leaf node, *Rule 2* outlines a relationship made of a connector plus a list of sub-relationships with root nodes in the list of the destination nodes of the connector, *Rule 3* is the termination for *Rule 4*, and *Rule 4* is needed to manage the junction of the disjoint sub-relationships with roots in the list $[X|Xs]$. When we compose connectors and tree-shaped relationships (*Rule 2* and *Rule 4*), we use the *times* clause to compose their trust/confidence values together.

To solve the search over the *and-or* graph problem it is enough to perform a query in Prolog language: for example, if we want to compute the cost of the best relationship rooted at n_1 (i.e. n_1 is the starting trustor) and having as leaves the

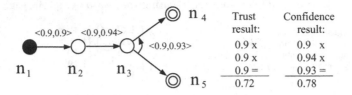

Fig. 2. The best trust relationship that can be found with the query $trust(n_1, [n_4, n_5], [T, C])$ for the program in Tab. 2

nodes representing the trustees (i.e. n_4 and n_5), we have to perform the query $trust(n_1, [n_4, n_5], [T, C])$, where T and C will be respectively instantiated with the trust and confidence values of the found relationship. The output for this query corresponds to the cost of the tree in Fig. 2, i.e. $\langle 0.72, 0.78 \rangle$. Otherwise, if we are interested in knowing the best trust relationship between one trustor (e.g. n_1) and only one trustee (e.g. n_4), as in classical trust propagation, we should perform the query $trust(n_1, [n_4], [T, C])$.

Notice that if the ratings of our trust relationships are objective (see Sec. 3), it is possible to directly program in CIAO also the ∘ operator explained in Sec. 3, and it would consequently be possible to build the n-connectors with $n > 1$ in the program, by applying the ∘ operator on the interested 1-connectors. In the program in Tab. 2 all the n-connectors are instead directly expressed as facts, and not automatically built with clauses.

5   An Implementation of the Model

To develop and test a practical implementation of our model, we adopt the *Java Universal Network/Graph Framework* [18], a software library for the modeling, analysis, and visualization of data that can be represented as a graph or network. The *WattsBetaSmallWorldGenerator* included in the library is a graph generator that produces a random small world network using the beta-model as proposed in [23]. The basic ideas is to start with a one-dimensional ring lattice in which each vertex has k-neighbors and then randomly rewire the edges, with probability β, in such a way that a small-world networks can be created for certain values of β and k that exhibit low characteristic path lengths and high clustering coefficient.

We generated the small-world network in Fig. 3 (with undirected edges) and then we automatically produced the corresponding program in CIAO (considering the edges as directed), as in Sec. 4. The relative statistics reported in Fig. 3 suggest the small-world nature of our test network: a quite high clustering coefficient and a low average shortest path.

With respect to the program in Tab. 2 we added the $Trust\_Hops < \lceil 2 \cdot Avg\_Shortest\_Path \rceil$ constraint: in this case, $Trust\_Hops < 9$, which is also the diameter of the network (see Fig. 3). This constraint limits the search space and provides a good approximation at the same time: in scale-free networks, the average distance between two nodes is logarithmic in the number of nodes [24],

Nodes	Edges	Clustering	Avg. SP	Min Deg	Max Deg.	Avg. Deg	Diameter
150	450	0.44	4.26	4	8	6	9

Fig. 3. The test small-world network generated with JUNG [18] and the corresponding statistics

i.e. every two nodes are close to each other. Therefore, this hop constraint can be successfully used also with large networks, and limiting the depth to twice the average shortest path value still results in a large number of alternative routes. We performed 50 tests on the graph in Fig. 3, and in every case the propagation between two nodes was computed within 5 minutes. Clearly, even if the results are promising and the small-world nature allows them to be repeated also on larger graph due to the logarithmic increase of average shortest path statistics, we need some improvements to further relax the *Trust_Hops* constraint. These improvements are suggested in Sec. 5.1.

5.1   Complexity Considerations

The representation of TN given in Sec. 3 can lead to an exponential time solution because of the degree of the nodes: for each of the individuals we have a connector towards each of the subsets of individuals in their social neighborhood, whose number is $O(2^d)$, where d is the out-degree of the node. The complexity of the tree search can be reduced by using *Tabled Constraint Logic Programming* (TCLP), i.e. with *tabling* (or *memoing*) techniques (for example, tabling efficiency is shown in [19]).

The calls to *tabled* predicates are stored in a searchable structure together with their proven instances, and subsequent identical calls can use the stored answers without repeating the computation. The work in [20] explains how to port *Constraint Handling Rules* (CHR) to XSB (acronym of *eXtended Stony Brook*), and in particular its focus is on technical issues related to the integration of CHR with tabled resolution. CHR is a high-level natural formalism to specify constraint solvers and propagation algorithm. At present time, from the XSB system it is possible to load a CHR package and to use its solving functionalities combined with tabling. This could be the promising framework where to solve QoS routing problems, since soft constraint satisfaction have already been successfully implemented in the CHR system [2].

The procedure of finding such a goal table for each single sub-community is much less time consuming than finding it for a whole not-partitioned social network. For this reason we can take advantage from the highly clustered nature of small-worlds. In Fig. 4 it is represented the community of people practising sports; the community is clustered into three sub-groups: Football, Basketball and Rugby. The individuals that represent the bridges among these groups are people practising two different sports, and are called *pivots*; their very important relationships are instead called *weak ties* (as we explained in Sec. 2), and can be used to widen the knowledge from a sub-group towards the rest of the small-world. If *Alice* (a pivot in the Basketball cluster) wants to retrieve a trust

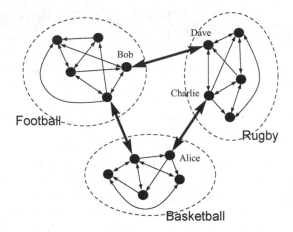

Fig. 4. The small-world of sports

score about *Bob* (a pivot in the Football cluster), she could ask to *Charlie* and *Charlie* to *Dave* (pivots in the Rugby cluster). Therefore, the pivots should store a "tabled vision" of their community to improve the performances for intra-community relationships.

6   Conclusions

We have defined the concept of multitrust and we have described a method to represent and solve the trust propagation problem with the combination of *and-or* graph and SCLP programming. Our framework can be fruitfully applied to have a quick and elegant formal-model where to compute the results of different trust metrics; in this paper we used trust and confidence, thus a precision estimation of the trust observation. We think that multitrust can be used in many real-world cases: trusting a group of individuals at the same time can lead to different conclusions w.r.t. simply aggregating together the trust values of the single individuals in the group. Then, we have provided a practical implementation of the model and we have tested it on a small-world social network, where all the individuals are reachable among themselves within few hops. The tests show that the framework can be used with small/medium networks with few hundreds of nodes due to small-world properties, but the performance need to be further improved. However, we provided many suggestions on how to reduce the complexity, and we will address these enhancements in future works.

A first improvement could be the use of memoization/tabling techniques, to filter out redundant computations. Then we plan also to extend SCLP in order to deal with *branch and bound* and to immediately prune the not promising partial solutions. These techniques can benefit from the small-world nature of the social networks, since the community is always partitioned in sub-groups with pivot individuals: we can have many small tables of goals instead of a big one for the whole network.

Our future goal is to find a structure able to aggregate distinct trust paths in a single trust value, i.e. to compute *multipath* propagation (e.g. an average cost of the independent paths). A solution could be represented by the expectation semiring [11], which is however somehow in contrast with pruning algorithms. At last, we would like to introduce the notion of "distrust" in the model and to propagate it by using the inverse of the semiring × operator [3].

References

1. Bistarelli, S. (ed.): Semirings for Soft Constraint Solving and Programming. LNCS, vol. 2962. Springer, Heidelberg (2004)
2. Bistarelli, S., Frühwirth, T., Marte, M.: Soft constraint propagation and solving in chrs. In: SAC 2002: Proc. of the 2002 ACM symposium on Applied computing, pp. 1–5. ACM Press, New York (2002)
3. Bistarelli, S., Gadducci, F.: Enhancing constraints manipulation in semiring-based formalisms. In: ECAI 2006, pp. 63–67. IOS Press, Amsterdam (2006)
4. Bistarelli, S., Montanari, U., Rossi, F.: Constraint solving over semirings. In: Proc. IJCAI 1995, pp. 624–630. Morgan Kaufman, San Francisco (1995)
5. Bistarelli, S., Montanari, U., Rossi, F.: Semiring-based constraint logic programming. In: Proc. IJCAI 1997, pp. 352–357. Morgan Kaufmann, San Francisco (1997)
6. Bistarelli, S., Montanari, U., Rossi, F.: Semiring-based constraint solving and optimization. Journal of the ACM 44(2), 201–236 (1997)
7. Bistarelli, S., Montanari, U., Rossi, F., Santini, F.: Modelling multicast QoS routing by using best-tree search in and-or graphs and soft constraint logic programming. Electr. Notes Theor. Comput. Sci. 190(3), 111–127 (2007)
8. Bistarelli, S., Santini, F.: Propagating multitrust within trust networks. In: SAC 2008, TRECK Track. ACM, New York (to appear, 2008)
9. Bueno, F., Cabeza, D., Carro, M., Hermenegildo, M., López-García, P., Puebla, G.: The CIAO prolog system: reference manual. Technical Report CLIP3/97.1, School of Computer Science, Technical University of Madrid (UPM) (1997)
10. Costa, A.C., Roe, R.A., Taillieu, T.: Trust within teams: the relation with performance effectiveness. European Journal of Work and Organizational Psychology 10(3), 225–244 (2001)
11. Eisner, J.: Parameter estimation for probabilistic finite-state transducers. In: ACL 2002, pp. 1–8. Association for Computational Linguistics (2001)
12. Georget, Y., Codognet, P.: Compiling semiring-based constraints with clp (FD, S). In: Maher, M.J., Puget, J.-F. (eds.) CP 1998. LNCS, vol. 1520, pp. 205–219. Springer, Heidelberg (1998)
13. Granovetter, M.S.: The Strength of Weak Ties. The American Journal of Sociology 78(6), 1360–1380 (1973)
14. Gray, E., Seigneur, J.M., Chen, Y., Jensen, C.D.: Trust propagation in small worlds. In: iTrust, pp. 239–254. Springer, Heidelberg (2003)
15. Jaffar, J., Maher, M.J.: Constraint logic programming: a survey. Journal of Logic Programming 19/20, 503–581 (1994)
16. Jøsang, A., Ismail, R., Boyd, C.: A survey of trust and reputation systems for online service provision. Decis. Support Syst. 43(2), 618–644 (2007)
17. Martelli, A., Montanari, U.: Optimizing decision trees through heuristically guided search. Commun. ACM 21(12), 1025–1039 (1978)
18. O'Madadhain, J., Fisher, D., White, S., Boey, Y.: The JUNG (Java Universal Network/Graph) framework. Technical report, UC Irvine (2003)

19. Ramakrishnan, I.V., Rao, P., Sagonas, K.F., Swift, T., Warren, D.S.: Efficient tabling mechanisms for logic programs. In: International Conference on Logic Programming, pp. 697–711. The MIT Press, Cambridge (1995)
20. Schrijvers, T., Warren, D.S.: Constraint handling rules and tabled execution. In: Demoen, B., Lifschitz, V. (eds.) ICLP 2004. LNCS, vol. 3132, pp. 120–136. Springer, Heidelberg (2004)
21. Theodorakopoulos, G., Baras, J.S.: Trust evaluation in ad-hoc networks. In: WiSe 2004: Workshop on Wireless security, pp. 1–10. ACM, New York (2004)
22. Twigg, A., Dimmock, N.: Attack-resistance of computational trust models. In: WETICE 2003, pp. 275–280. IEEE Computer Society Press, Los Alamitos (2003)
23. Watts, D.J.: Small worlds: the dynamics of networks between order and randomness. Princeton University Press, Princeton (1999)
24. Watts, D.J., Strogatz, S.H.: Collective dynamics of small-world networks. Nature 393, 440 (1998)
25. Ziegler, C.N., Lausen, G.: Propagation models for trust and distrust in social networks. Information Systems Frontiers 7(4-5), 337–358 (2005)

Improving ABT Performance by Adding Synchronization Points*

Ismel Brito and Pedro Meseguer

IIIA, Institut d'Investigació en Intel.ligència Artificial
CSIC, Consejo Superior de Investigaciones Científicas
Campus UAB, 08193 Bellaterra, Spain
{ismel,pedro}@iiia.csic.es

Abstract. Asynchronous Backtracking (*ABT*) is a reference algorithm for Distributed CSP (*DisCSP*). In *ABT*, agents assign values to their variables and exchange messages asynchronously and concurrently. When an *ABT* agent sends a backtracking message, it continues working without waiting for an answer. In this paper, we describe a case showing that this strategy may cause some inefficiency. To overcome this, we propose ABT_{hyb}, a new algorithm that results from adding synchronization points to *ABT*. We prove that ABT_{hyb} is correct, complete and terminates. We also provide an empirical evaluation of the new algorithm on several benchmarks. Experimental results show that ABT_{hyb} outperforms *ABT*.

1 Introduction

In recent years, there is an increasing interest for solving problems in which information is distributed among different agents. In standard Constraint Satisfaction Problems (*CSP*), centralized solving is assumed, so it is inadequate for problems requiring a true distributed resolution. This has motivated the new Distributed CSP (*DisCSP*) framework, where constraint problems with elements (variables, domains, constraints) distributed among automated agents which cannot be centralized for different reasons (prohibitive translation costs or security/privacy issues) are modelled and solved.

When solving *DisCSP*, all agents cooperate for finding a globally consistent solution, that is an assignment involving all variables that satisfies every constraint. To achieve this, agents assign their variables and exchange messages on these assignments, which allows them to check their consistency with respect to problem constraints. Several synchronous and asynchronous procedures have been proposed [2,3,6,7,8].

Broadly speaking, a synchronous algorithm is based on the notion of *privilege*, a token that is passed among agents. Only the agent that has the privilege is active, while the rest of agents are waiting [1]. When the active agent terminates, it passes the privilege to another agent, which now becomes the active one. These algorithms have a low degree of parallelism, but agents receive up to date information. Alternatively, asynchronous algorithms allow agents to be active concurrently. Agents may assign their variables

---

* Supported by the Spanish project TIN2006-15387-C03-01.
[1] Except for special topological arrangements of the constraint graph. See [3] for a synchronous algorithm where several agents are active concurrently.

F. Fages, F. Rossi, and S. Soliman (Eds.): CSCLP 2007, LNAI 5129, pp. 47–61, 2008.
© Springer-Verlag Berlin Heidelberg 2008

and exchange messages asynchronously. These algorithms show a high degree of parallelism, but the information that an agent knows about other agents is less up to date than in synchronous procedures.

In this paper we study the effect of adding synchronization points to Asynchronous Backtracking (*ABT*), a reference algorithm for *DisCSP* [9]. In *ABT*, agents assign values to their variables and exchange messages asynchronously and concurrently. When an *ABT* agent sends a backtracking message, it continues working without waiting for an answer. This strategy may be costly in some cases because performing two tasks concurrently could be inefficient if there is a dependency relation between them. We identify a case in which *ABT*'s efficiency can be improved if, after backtracking, an agent waits for receiving a message showing the effect of backtracking on higher priority agents. We implement this idea on ABT_{hyb}, a new *ABT*-like algorithm, that combines asynchronous and synchronous elements to avoid redundant messages. We show that ABT_{hyb} is correct, complete and terminates. Experimental results indicate that ABT_{hyb} outperforms *ABT* on three different benchmarks.

This paper is organized as follows. First, we recall the definition of *DisCSP*, we describe *ABT* and identify a source of inefficiency in it. Then, we present ABT_{hyb}, a new hybrid algorithm that combines asynchronous and synchronous elements. We prove some properties of the new algorithm, which is evaluated on three different benchmarks. Finally, we extract some conclusions from this work.

2 Preliminaries

A *Constraint Satisfaction Problem* $(\mathcal{X}, \mathcal{D}, \mathcal{C})$ involves a finite set of variables $\mathcal{X}$, each taking values in a finite domain, and a finite set of constraints $\mathcal{C}$. A constraint on a subset of variables forbids some combinations of values that these variables can take. A *solution* is an assignment of values to variables which satisfies every constraint. Formally,

- $\mathcal{X} = \{x_1, \ldots, x_n\}$ is a set of n variables;
- $\mathcal{D} = \{D(x_1), \ldots, D(x_n)\}$ is a collection of finite domains; $D(x_i)$ is the initial set of possible values for x_i;
- $\mathcal{C}$ is a finite set of constraints. A constraint C_i on the ordered subset of variables $var(C_i) = (x_{i_1}, \ldots, x_{i_{r(i)}})$ specifies the relation $prm(C_i)$ of the *permitted* combinations of values for the variables in $var(C_i)$. Clearly, $prm(C_i) \subset \prod_{x_{i_k} \in var(C_i)} D(x_{i_k})$. An element of $prm(C_i)$ is a tuple $(v_{i_1}, \ldots, v_{i_{r(i)}})$, $v_{i_k} \in D(x_{i_k})$.

A *Distributed Constraint Satisfaction Problem* (*DisCSP*) is a *CSP* where variables, domains and constraints are distributed among automated agents. Formally, a finite *DisCSP* is defined by a 5-tuple $(\mathcal{X}, \mathcal{D}, \mathcal{C}, \mathcal{A}, \phi)$, where $\mathcal{X}$, $\mathcal{D}$ and $\mathcal{C}$ are as before, and

- $\mathcal{A} = \{1, \ldots, p\}$ is a set of p agents,
- $\phi : \mathcal{X} \to \mathcal{A}$ is a function that maps each variable to its agent.

Each variable belongs to one agent. The distribution of variables divides $\mathcal{C}$ in two disjoint subsets, $\mathcal{C}_{intra} = \{C_i | \forall x_j, x_k \in var(C_i), \phi(x_j) = \phi(x_k)\}$, and $\mathcal{C}_{inter} = \{C_i | \exists x_j, x_k \in var(C_i), \phi(x_j) \neq \phi(x_k)\}$, called intraagent and interagent constraint sets, respectively. An intraagent constraint C_i is known by the agent owner of $var(C_i)$, and it is unknown by the other agents. Usually, it is considered that an interagent constraint C_j is known by every agent that owns a variable of $var(C_j)$ [9].

A *solution* of a *DisCSP* is an assignment of values to variables satisfying every constraint. *DisCSP*s are solved by the coordinated action of agents, which communicate by exchanging messages. It is assumed that the delay of a message is finite but random. For a given pair of agents, messages are delivered in the order they were sent.

For simplicity purposes, and to emphasize the distribution aspects, we assume that each agent owns exactly one variable. We identify the agent number with its variable index ($\forall x_i \in \mathcal{X}, \phi(x_i) = i$). From this assumption, all constraints are interagent constraints, so $C = C_{inter}$ and $C_{intra} = \emptyset$. Furthermore, we assume that all constraints are binary. A constraint is written C_{ij} to indicate that it binds variables x_i and x_j.

3   The Asynchronous Backtracking Algorithm

Asynchronous backtracking (*ABT*) [7,9] was a pioneering algorithm for *DisCSP* solving, its first version dates from 1992. *ABT* is an asynchronous algorithm, that allows agents to act asynchronous and concurrently. An *ABT* agent makes its own decisions, informs other agents about them, and no agent has to wait for the others' decisions. The algorithm computes a global consistent solution (or detects that no solution exists) in finite time; its correctness and completeness have been proved [9,2]. *ABT* requires constraints to be directed. A binary constraint causes a directed link between the two constrained agents: the value-sending agent, from which the link starts, and the constraint-evaluating agent, at which the link ends. To make the network cycle-free, there is a total order among agents, which is followed by the directed links.

The *ABT* algorithm is executed by each agent, which keeps its own agent view and nogood store. Considering a generic agent $self$, its agent view is the set of values that $self$ believes are assigned to higher priority agents (connected to $self$ by incoming links). Its nogood store keeps nogoods received by $self$ as justifications of inconsistent values. *ABT* agents exchange three types of messages: **ok?** (assignments), **ngd** (nogoods) and **adl** (link request). An **stp** message indicates that there is no solution.

When the algorithm starts, each agent assigns its variable, and sends the assignment to its neighboring agents with lower priority. When $self$ receives an assignment, $self$ updates its agent view with the new assignment, removes inconsistent nogoods and checks the consistency of its current assignment with the updated agent view.

When $self$ receives a nogood, it is accepted if the nogood is consistent with $self$'s agent view (for the variables in the nogood, their values in the nogood and in $self$'s agent view are equal). Otherwise, $self$ discards the nogood as obsolete. If the nogood is accepted, the nogood store is updated, causing $self$ to search for a new consistent value (since the received nogood forbids its current value.) When an agent cannot find any value consistent with its agent view, either because of the original constraints or because of the received nogoods, new nogoods are generated from its agent view and each one sent to the closest agent involved in it. This operation causes backtracking.

There are several versions of *ABT*, depending on how new nogoods are generated. The simplest form is to send the complete agent view as nogood [10]. In [2], when an agent has no consistent values, it resolves its nogoods following a procedure described in [1]. In this paper we consider this last version.

The majority of messages exchanged by *ABT* agents are **ok?** and **ngd**. While **ok?** messages are always accepted, some **ngd** messages may be discarded as obsolete. *ABT* could save some work if these discarded messages were not sent. Although the sender agent cannot detect which messages will become obsolete when reaching the receiver, it is possible to avoid sending those which are redundant.

When agent j sends a **ngd** message, it performs a new assignment and informs lower priority agents, without waiting for the reception of any message showing the effect of that **ngd** on higher priority agents. This behavior can generate inefficiency in the following situation. If k sends a **ngd** message to j causing a domain wipe-out in j, then j sends a **ngd** message to some previous agent i, deleting the value of i in its agent view. If j takes the same value as before and sends an **ok?** message to k before receiving the new value for i, k will find again the same inconsistency so it will send the same nogood to j in a new **ngd** message. Since j has forgotten i's value, j will discard the **ngd** message as obsolete, sending again its value to k in an **ok?** message. The process is repeated generating useless messages, until some higher variable changes its value and the corresponding **ok?** message arrives at j and k. In the next section we propose a new *ABT* version to avoid sending these redundant messages.

4   The ABT_{hyb} Algorithm

To avoid the behavior described above, we present ABT_{hyb}, a hybrid algorithm that combines asynchronous and synchronous elements. ABT_{hyb} behaves like *ABT* when no backtracking is performed: agents take their values asynchronously and inform lower priority agents. However, when an agent has to backtrack, it does it synchronously as follows. If k has no value consistent with its agent view, it sends a **ngd** message to j and enters a *waiting* state. In this state, k has no assigned value, and it does not send out any message. Any received **ok?** message is accepted, updating k's agent view accordingly. Any received **ngd** message is treated as obsolete, since k has no value assigned. Agent k leaves the waiting state when receiving one the following messages:

1. An **ok?** message that breaks the nogood sent by k.
2. An **ok?** message from j, the receiver of the last **ngd** message.
3. A **stp** message informing that the problem has not solution.

The justification for leaving the waiting state is as follows,

1. An **ok?** message breaking the nogood confirms that the **ngd** message has caused a change in a higher priority agent such that the system escapes from the nogood. Agent k has to leave the waiting state, returning to ordinary *ABT* operation.
2. An **ok?** message from j, considers the situation in which k has a more updated information than j when k sends the **ngd** and enters the waiting state. While j does not receive the updated information, it will reject the **ngd** message as obsolete and resend its value to k in an **ok?** message. After receiving it, if k remains in the waiting state the communication with j might be broken, because j may say nothing

when receiving the updated information, k will have no notice of this updated information and the algorithm would be incomplete. So k has to leave the waiting state, just to rediscover the same nogood, send it to j and enter the waiting state again. This loop breaks when the updated information reaches j: it will no longer reject the **ngd** because it will not be obsolete according to its updated agent view.

3. A **stp** message indicates that the empty nogood has been generated somewhere, so every agent has to finish its execution.

At this point, ABT_{hyb} switches to ABT. ABT_{hyb} detects that a $DisCSP$ is unsolvable if during the resolution an empty nogood is derived. Otherwise, ABT_{hyb} claims that it has found a solution when no messages are traveling through the network (quiescence).

The ABT_{hyb} algorithm appears in Figure 1. The sets Γ^- and Γ^+ are formed by higher and lower priority agents, respectively. A nogood ng is written in its ordered form, $ng : x_1 = a \land x_2 = b \Rightarrow x_3 \neq c$. The conjunction at the left-hand side of $\Rightarrow$ is written as $\mathtt{lhs}(ng)$, while the assignment at the right-hand side is written as $\mathtt{rhs}(ng)$.

The differences between ABT_{hyb} and ABT [2] could be seen following variable $wait$. When the agent initiates a backtracking, sending a **ngd** message, this nogood is memorized in $lastNogood$ and $wait$ takes value true (line 6 of $\mathtt{Backtrack}$). From this point on, the agent is in the waiting state, in which it accepts any **ok?** message, updating its agent view (line 1 of $\mathtt{ProcessInfo}$), but it discards any **ngd** message (line 7 of $\mathtt{ABT-Hyb}$). Since the agent has no value, no **ok?** message departs from it (line 5 of $\mathtt{ProcessInfo}$ and line 2 of $\mathtt{SetLink}$). Since the agent does not try to get a new value, no **ngd** message departs from it (line 5 of $\mathtt{ProcessInfo}$). The agent can leave the waiting state after (1) receiving an **ok?** message breaking $lastNogood$, (2) receiving an **ok?** message from the agent destination of the **ngd** message (lines 3 and 4 of $\mathtt{ProcessInfo}$), or (3) receiving a **stp** message (line 8 of $\mathtt{ABT-Hyb}$).

4.1 Correctness, Completeness and Termination

Regardless of synchronous points, ABT_{hyb} inherits the good properties of ABT, namely correctness, completeness and termination. To prove them, we start with some lemmas.

Lemma 1. *No ABT_{hyb} agent will continue in a waiting state forever.*

Proof. (By induction). In ABT_{hyb} an agent enters the waiting state after sending a **ngd** message to a higher priority agent. The first agent in the ordering will not enter the waiting state because no **ngd** message departs from it. Suppose that no agent in $1, 2, \ldots, k-1$ is waiting forever, and suppose that agent k enters the waiting state after sending a **ngd** message to j ($1 \leq j \leq k-1$). We will show that k will not be forever in the waiting state. When j receives the **ngd** message, there are two possible states:

1. j is waiting. Since no agent in $1, 2, \ldots, k-1$ is waiting forever, j will leave the waiting state at some point. If x_j has a value consistent with its new agent view, j will send an **ok?** message to k. If x_j has no value consistent with its new agent view, j will backtrack and enter again the waiting state. This can be done a finite number of times (because there is a finite number of values per variable) before finding a consistent value or discovering that the problem has no solution (generating a **stp** message). In both cases, agent k will leave the waiting state.

```
procedure ABT-Hyb()
1 myValue ← empty; end ← false; wait ← false;
2 CheckAgentView();
3 while (¬end) do
4   msg ← getMsg();
5   switch(msg.type)
6     ok?  : ProcessInfo(msg);
7     ngd  : if ¬wait then ResolveConflict(msg);
8     stp  : wait ← false; end ← true;
9     adl  : SetLink(msg);

procedure CheckAgentView(msg)
1 if ¬consistent(myValue, myAgentView) then
2   myValue ← ChooseValue();
3   if (myValue) then for each child ∈ Γ⁺(self) do sendMsg:ok?(child, myValue);
4   else Backtrack();

procedure ProcessInfo(msg)
1 Update(myAgentView, msg.Assig);
2 if wait then
3   if (msg.Sender ∈ rhs(lastNogood)) ∨ (msg.Sender ∈ lhs(lastNogood) ∧ msg.Assig ≠ lastNogood[msg.Sender])
4     then wait ← false;
5 if ¬wait then CheckAgentView();

procedure ResolveConflict(msg)
1   if Coherent(msg.Nogood, Γ⁻(self) ∪ {self}) then
2     CheckAddLink(msg);
3     add(msg.Nogood, myNogoodStore); myValue ← empty;
4     CheckAgentView();
5   else if Coherent(msg.Nogood, self) then sendMsg:ok?(msg.Sender, myValue);

procedure Backtrack()
1 newNogood ← solve(myNogoodStore);
2 if (newNogood = empty) then
3   end ← true; sendMsg:Stop(system);
4 else
5   sendMsg:ngd(newNogood);
6   lastNogood ← newNogood; wait ← true;

function ChooseValue()
1 for each v ∈ D(self) not eliminated by myNogoodStore do
2   if consistent(v, myAgentView) then return (v);
3   else add(xⱼ = valⱼ ⇒ self ≠ v, myNogoodStore); /*v is inconsistent with xⱼ's value */
4 return (empty);

procedure Update(myAgentView, newAssig)
1 add(newAssig, myAgentView);
2 for each ng ∈ myNogoodStore do
3   if ¬Coherent(lhs(ng), myAgentView) then remove(ng, myNogoodStore);

function Coherent(nogood, agents)
1 for each var ∈ lhs(nogood) ∪ agents do
2   if nogood[var] ≠ myAgentView[var] then return false;
3 return true;

procedure SetLink(msg)
1 add(msg.sender, Γ⁺(self));
2 if ¬wait then sendMsg:ok?(msg.sender, myValue);

procedure CheckAddLink(msg)
1 for each (var ∈ lhs(msg.Nogood))
2   if (var ∉ Γ⁻(self)) then
3     sendMsg:addl(var, self);
4     add(var, Γ⁻(self));
5     Update(myAgentView, var ← varValue);
```

Fig. 1. The ABT$_{hyb}$ algorithm for asynchronous backtracking search

2. j is not waiting. The **ngd** message could be:
   (a) Obsolete in the value of x_j. In this case, there is an **ok?** message traveling from j to k that has not arrived at k yet. After receiving such a message, k will leave the waiting state.
   (b) Obsolete not in the value of x_j. In this case, j resends x_j value to k in an **ok?** message. After receiving such a message, k will leave the waiting state.
   (c) Not obsolete. The current value of x_j is forbidden by the received nogood, so a new value is tried. If j finds another value consistent with its agent view, x_j takes it and sends an **ok?** message to k, which leaves the waiting state. Otherwise, j backtracks to a previous agent in the ordering, and enters the waiting state. Since no agent in $1, 2, \ldots, k-1$ is waiting forever, j will leave the waiting state at some point. As explained in the point 1 above, this causes that k leaves the waiting state as well.

Therefore, we conclude that agent k will not continue in a waiting state forever.   □

Lemma 2. *If an ABT$_{hyb}$ agent is in a waiting state, the network is not quiescent.*

Proof. An agent is in a waiting state after sending a **ngd** message. Because Lemma 1, this agent will leave the waiting state in finite time. This is done after receiving an **ok?** or **stp** message. Therefore, if there is an agent in a waiting state, the network cannot be quiescent at least until one of those messages has been produced.   □

Lemma 3. *In ABT$_{hyb}$, a nogood that is discarded as obsolete because the receiver is in a waiting state, will be resent to the receiver until the sender realizes that it has been solved, or the empty nogood has been derived.*

Proof. If agent k sends a nogood to agent j that is in a waiting state, this nogood is discarded and k enters the waiting state. From Lemma 1, no agent will continue in a waiting state forever, so k will leave that state in finite time. This is done after receiving,

- An **ok?** message from j. If this message does not solve the nogood, it will be generated and resend to j. If it solves it, this nogood is not generated, exactly in the same way as *ABT* does.
- An **ok?** message from an agent higher than j, breaking the nogood. Since the nogood is no longer active, it is not resent again.
- A **stp** message. The process terminates without solution.

So the nogood is sent again until it is solved or the empty nogood is generated.   □

Proposition 1. *ABT$_{hyb}$ is correct.*

Proof. From Lemma 2, *ABT$_{hyb}$* reaches quiescence when no agent is in a waiting state. From this fact, *ABT$_{hyb}$* correctness derives directly from *ABT* correctness: when the network is quiescent all agents satisfy their constraints, so their current assignments form a solution. If this would not be the case, at least one agent would detect a violated constraint and it would send a message, breaking the quiescence assumption.   □

Proposition 2. *ABT$_{hyb}$ is complete and terminates.*

Proof. From Lemma 3, the synchronicity of backtracking in ABT_{hyb} does not cause to ignore any nogood. Then, ABT_{hyb} explores the search space as ABT does. From this fact, ABT_{hyb} completeness comes directly from ABT completeness. New nogoods are generated by logical inference, so the empty nogood cannot be derived if there is a solution. Total agent ordering causes that backtracking discards one value in the highest variable reached by a **ngd** message. Since the number of values is finite, the process will find a solution if it exists, or it will derive the empty nogood otherwise.

To see that ABT_{hyb} terminates, we have to prove that no agent falls into an infinite loop. This comes from the fact that agents cannot continue in a waiting state forever (Lemma 1), and that ABT agents cannot be in an endless loop.   □

Instead of adding synchronization points, we can avoid resending redundant **ngd** messages with exponential-space algorithms. Let us assume that $self$ stores every nogood sent, while it is not obsolete. When a domain wipe-out occurs in $self$, if the new generated nogood is equal to one of the stored nogoods, it is not sent. This avoids $self$ sending identical nogoods until some higher agent changes its value and the corresponding **ok?** arrives at $self$. But this requires exponential space, since the number of nogoods generated could be exponential in the number of agents with higher priority. A similar idea appears in [8] for the asynchronous weak-commitment algorithm.

4.2   Comparison with ABT

In practice, ABT_{hyb} shows a better performance than ABT. However, ABT_{hyb} does not always require less messages than ABT for solving a particular instance. In the following we provide an example of this. Let us consider the instance depicted in Figure 2. It has 7 agents, each holding a variable, their domains are indicated, and with equality and disequality constraints. For ABT and ABT_{hyb}, we assume the same network conditions (i.e. messages are received in the same order for both algorithms), the same agent ordering (lexicographical, from A_1 to A_7) and the same value ordering (values are tried in the order they appear in Figure 2). Initially, each agent assigns its variable with the first value of its domain and sends the corresponding **ok?** message to its descendants. When all arrive, except the message coming from A_1, the assignment, agent view and nogood store of every agent appear in Figure 3. The situation is the same for both algorithms.

At this point, A_7 has a conflict (referred as α): no value of x_7 is consistent with its agent view, so A_7 backtracks (it resolves its nogood store and generates a new nogood, which is sent to the closest agent involved). After this decision, ABT and ABT_{hyb} behave differently. ABT behavior is summarized in Figure 4. In step 1, A_7 sends a **ngd** message to A_6, implementing the backtracking initiated by A_7. This message will arrive at its

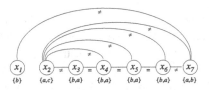

Fig. 2. Instance with 7 agents, each holding a variable. Domains and constraints are indicated.

A_1	A_2	A_3	A_4	A_5	A_6	A_7
$x_1 \leftarrow b$	$x_2 \leftarrow a$	$x_3 \leftarrow b$	$x_4 \leftarrow b$	$x_5 \leftarrow b$	$x_6 \leftarrow b$	$x_7 \leftarrow$ no value
		$x_2 = a$	$x_2 = a, x_3 = b$	$x_2 = a, x_4 = b$	$x_2 = a, x_5 = b$	$x_2 = a, x_6 = b$
						$x_2 = a \Rightarrow x_7 \neq a$
						$x_6 = b \Rightarrow x_7 \neq b$

Fig. 3. First assignments, agent views and nogood stores of the agents after receiving the corresponding **ok?** messages (except the **ok?** coming from A_1). Same situation for ABT and ABT_{hyb}.

step	#msg	type	from	to	message	comments
1	1	**ngd**	A_7	A_6	$x_2 = a \Rightarrow x_6 \neq b$	resolution of A_7 nogood store this message arrives at A_6 after step 4 $x_7 \leftarrow b$
2	1	**ok?**	A_1	A_7	$x_1 = b$	x_7 looks for a compatible value generates nogood $x_1 = b \Rightarrow x_7 \neq b$ resolution with $x_2 = a \Rightarrow x_7 \neq a$ causes new nogood of step 3
3	1	**ngd**	A_7	A_2	$x_1 = b \Rightarrow x_2 \neq a$	causes change in x_2
4	5	**ok?**	A_2	$A_3 - A_7$	$x_2 = c$	new value of x_2
5	1	**ok?**	A_6	A_7	$x_6 = b$	message of step 1 arrived at A_6 discarded as obsolete, because now $x_2 = c$ value of x_6 is resent to A_7 a solution b, c, b, b, b, b, a is found

Fig. 4. Messages exchanged by ABT agents just after the situation depicted in Figure 3

step	#msg	type	from	to	message	comments
1	1	**ngd**	A_7	A_6	$x_2 = a \Rightarrow x_6 \neq b$	generated by resolution of A_7 nogood store, A_7 is waiting add to A_6 nogood store $x_2 = a \Rightarrow x_6 \neq b$ x_6 looks for a compatible value add to A_6 nogood store $x_5 = b \Rightarrow x_6 \neq a$
2	1	**ok?**	A_1	A_7	$x_1 = b$	updates A_7 agent view $x_1 = b$
3	1	**ngd**	A_6	A_5	$x_2 = a \Rightarrow x_5 \neq b$	generated by resolution of A_6 nogood store, A_6 is waiting add to A_5 nogood store $x_2 = a \Rightarrow x_5 \neq b$ x_5 looks for a compatible value add to A_5 nogood store $x_4 = b \Rightarrow x_5 \neq a$
4	1	**ngd**	A_5	A_4	$x_2 = a \Rightarrow x_4 \neq b$	generated by resolution of A_5 nogood store, A_5 is waiting add to A_4 nogood store $x_2 = a \Rightarrow x_4 \neq b$ x_4 looks for a compatible value add to A_4 nogood store $x_3 = b \Rightarrow x_4 \neq a$
5	1	**ngd**	A_4	A_3	$x_2 = a \Rightarrow x_3 \neq b$	generated by resolution of A_4 nogood store, A_4 is waiting add to A_3 nogood store $x_2 = a \Rightarrow x_3 \neq b$ x_4 looks for a compatible value add to A_3 nogood store $x_2 = a \Rightarrow x_3 \neq a$
6	1	**ngd**	A_3	A_2	$\Rightarrow x_2 \neq a$	generated by resolution of A_3 nogood store, A_3 is waiting add to A_2 nogood store $\Rightarrow x_2 \neq a, x_2 \leftarrow c$
7	5	**ok?**	A_2	$A_3 - A_7$	$x_2 = c$	new value of x_2 agents $A_3 - A_7$ wake up a solution b, c, b, b, b, b, a is found

Fig. 5. Messages exchanged by ABT_{hyb} agents just after the situation depicted in Figure 3

destination after step 4. In step 2, the initial **ok?** message coming from A_1 arrives at A_7. This message causes again another conflict (referred as β) in A_7, so it initiates a new backtracking. In step 3, A_7 sends a new **ngd** message to A_2. This message causes A_2 to change the value of x_2 to c. A_2 informs of this new assignment to $A_3 - A_7$ in step 4, using 5 **ok?** messages. After this, the **ngd** of step 1 arrives at A_6. This message is discarded as obsolete, because the new value of x_2. In step 5, the value of x_6 is resent to A_7. Solution b, c, b, b, b, b, a is found, requiring 9 messages since the first backtracking.

ABT_{hyb} behavior is summarized in Figure 5. In step 1, A_7 sends a **ngd** message to A_6, implementing the backtracking initiated by A_7 as in the ABT case. Then, A_7 enters the waiting state. In step 2, the initial **ok?** message coming from A_1 arrives at A_7. Since A_7 is waiting, it just updates its agent view, without any other action. In step 3, nogood sent in step 1 arrives at A_6. This causes that x_6 has no value consistent with its agent view. A_6 resolves its nogood store and generates a new nogood which is sent to A_5. A_6 enters the waiting state. In steps 4, 5 and 6, the same happens to agents A_5, A_4, and A_3, a nogood is propagated backwards and these agents remain waiting. In step 7, the nogood sent in step 6 arrives at A_2, causing to change the value of x_2, which now takes c. In step 8, A_2 informs about the new assignment of x_2 with 5 **ok?** messages, to $A_3 - A_7$. Solution b, c, b, b, b, b, a is found, requiring 11 messages since the first backtracking. In this case, ABT needs less messages than ABT_{hyb}.

We have to solve two conflicts, α and β; α is discovered first and β is discovered before α is solved. Both are solved by the same action (changing x_2 to c). ABT may start the resolution of both conflicts in parallel, while ABT_{hyb} has to solve them sequentially. But solving β requires less messages than solving α, so ABT may need less messages than ABT_{hyb} to find a solution. However, if after discovering a conflict, no second conflict is found among the involved variables, the number of messages required by ABT_{hyb} is lower than (or equal to) the number required by ABT, as proved next.

Proposition 3. *The number of messages required by ABT_{hyb} to solve a conflict is not higher than the number of messages required by ABT to solve the same conflict, if during the resolution of this conflict no other conflict is found among conflicting variables.*

Proof. Let j and i be the agents that find the conflict and resolve it, respectively ($i < j$). Let S be the set of agents $\{i + 1 \ldots j\}$. We differentiate between **ok?** and **ngd**.

Regarding **ngd** messages, before i receives the nogood that causes to change the value of its variable, agents in S have exchanged **ngd** messages. The number of the **ngd** messages not discarded as obsolete (really contributing to the change of x_i) is the same in both algorithms. The number of obsolete nogoods because the receiver has a more updated information than the sender is the same in both algorithms. In addition, ABT may have redundant obsolete **ngd** (as explained in section 4), which cannot occur in ABT_{hyb}. So the number of **ngd** messages in ABT_{hyb} is no higher than in ABT.

Regarding **ok?** messages, each ABT_{hyb} agent in S will leave the waiting state after receiving one of the following messages: (1) an **ok?** from i breaking the nogood, (2) an **ok?** from any other higher priority agent breaking the nogood or (3) a **stp** message. Next, we count the number of messages sent by k in these cases:

(1) In ABT_{hyb}, x_k takes a value consistent with the new value of x_i and sends 1 **ok?** message to each lower priority agent. In ABT, after backtracking, x_k has to take a value without knowing the new value of x_i. This value may be consistent or inconsistent with the new value of x_i. If it is consistent, k will send 1 **ok?** message to each lower priority agent. Therefore, the number of messages sent by k in both algorithms is the same. If the value of x_k is inconsistent with the new value of x_i, k will send 2 **ok?** messages to each lower priority agent: one message because the inconsistent value and the other after assigning a value consistent with x_i. So the number of messages sent by k in ABT_{hyb} is not higher than the number sent in ABT.

(2) In ABT_{hyb}, x_k takes a new value before receiving the new assignment of x_i. This new value may be consistent or inconsistent. If the new value of x_k is consistent, then k will send 1 **ok?** message to each lower priority agent. Otherwise, it will send 2 **ok?** messages to each lower priority agent. This is the same situation that happens in ABT (see previous case). So the number of messages sent by k is the same in both algorithms. (3) In both algorithms when agent k receives a **stp** message, k does not send any more messages. Thus, the number of messages sent by k in both algorithms is the same.

So ABT_{hyb} never requires more messages than those required by ABT.   □

5 Experimental Results

We have evaluated ABT and ABT_{hyb} on three benchmarks: distributed n-queens, random binary $DisCSPs$ and Distributed Sensor-Mobile. We compare both algorithms on the computation effort, as the number of non concurrent constraint checks ($nccc$, usually used instead of computation time in distributed constraint solving [5]), and the communication cost, as the total number of messages exchanged (msg). We have used a discrete simulator to execute both algorithms. It randomly activates one agent at a time, which reads all incoming messages and processes them [2], sending new messages if needed. When the active agent terminates, a new agent is activated. If an agent detects several justifications for a forbidden value, it follows the strategy of selecting the *best nogood* (the nogood with the highest possible lowest agent involved) [2].

To measure the impact of network traffic conditions on the algorithmic performance, we have also evaluated the algorithms introducing random delays in message delivery. Delays are taken from a uniform distribution between 0 and 100 time units.

5.1 Distributed n-Queens

The distributed n-queens problem is the classical n-queens problem where each queen is hold by an independent agent. We ran ABT and ABT_{hyb} for 4 dimensions of this problem, n = 10, 15, 20, 25. Table 1 shows the results in terms of $nccc$ and msg, averaged over 100 executions. We observe that ABT_{hyb} is clearly better than ABT.

Table 2 reports the number of messages per type for executions without delays. We observe that ABT_{hyb} outperforms ABT for each type. The number of obsolete **ngd** messages in ABT_{hyb} decreases one order of magnitude with respect to the same type of messages in ABT. However, the global improvement goes beyond the savings in obsolete **ngd**, because **ok?** and **ngd** messages also decrease to a larger extent. This is due to the following collective effect. When an ABT agent sends a **ngd**, it tries to get a new consistent value without knowing the effect that backtracking causes in higher priority agents. If it finds such a consistent value, it informs to lower priority agents using **ok?** messages. If it happens that this value is not consistent with new values that backtracking causes in higher priority agents, these **ok?** would be useless, and new **ngd** would be generated. ABT_{hyb} tries to avoid this situation. When an ABT_{hyb} agent sends a **ngd** message, it waits until it receives notice of the effect of backtracking in higher priority

---

[2] In both algorithms agents process messages by packets, instead of one by one, because performance improves [11].

Table 1. Number of non concurrent constraint checks and messages required by ABT and ABT_{hyb} to solve instances of the distributed n-queens, without (left) and with (right) message delays

	ABT		ABT_{hyb}		d-ABT		d-ABT_{hyb}	
n	$nccc$	msg	$nccc$	msg	$nccc$	msg	$nccc$	msg
10	2,223	739	1,699	502	2,117	907	1,601	614
15	56,412	13,978	32,373	6,881	50,163	17,201	30,891	9,611
20	11,084,012	2,198,304	6,086,376	995,902	9,616,876	2,695,008	5,135,193	1,403,472
25	3,868,136	693,832	1,660,448	271,092	3,206,581	1,421,903	1,580,132	777,187

Table 2. The number of messages per type exchanged by ABT and ABT_{hyb} to solve instances of the distributed n-queens problem without message delays

	ABT			ABT_{hyb}		
n	ok?	ngd	obs ngd (%)	ok?	ngd	obs ngd (%)
10	546	193	67 (34.7%)	409	93	7 (7.5%)
15	10,029	3,949	1,515 (38.4%)	5,547	1,334	111 (8.3%)
20	1,609,727	588,578	239,763 (40.7%)	817,304	178,598	15,354 (8.6%)
25	518,719	175,113	76,771 (43.8%)	229,159	41,934	4,381 (10.4%)

agents. When it leaves the waiting state, it tries to get a consistent value. At this point, it knows some effect of the backtracking on higher priority agents, so the new value will be consistent with it. The new value has more chance to be consistent with higher priority agents, and its **ok?** messages will be more likely to make useful work.

In Table 1 (right) we have results for ABT and ABT_{hyb} with delays. Again, ABT_{hyb} is substantially better than ABT. We observe that the presence of delays degrades performance of both algorithms. ABT_{hyb} deteriorates slightly more than ABT. This behavior can be explained using the results of Table 2. With delays, the number of obsolete **ngd** (useless messages) in ABT_{hyb} increases, degrading the collective effect previously mentioned. Anyway, ABT_{hyb} remains substantially better than ABT in presence of delays.

Table 3. The number of messages per type exchanged by ABT and ABT_{hyb} to solve instances of the distributed n-queens problem with message delays

	d-ABT			d-ABT_{hyb}		
n	ok?	ngd	obs ngd (%)	ok?	ngd	obs ngd (%)
10	663	244	129 (52.8%)	478	136	29 (21.5%)
15	11,972	5,231	2,925 (55.9%)	7,902	1,709	496 (29.0%)
20	1,908,769	786,239	461,285 (58.7%)	1,155,783	247,689	107,992 (43.6%)
25	1,421,904	555,237	365,901 (65.9%)	631,752	145,435	76,208 (52.4%)

5.2   Random Problems

Uniform binary random CSPs are characterized by $\langle n, d, p_1, p_2 \rangle$ where n is the number of variables, d is the number of values per variable, p_1 is the network *connectivity* defined as the ratio of existing constraints, and p_2 is the constraint *tightness* defined as the ratio of forbidden value pairs. We generated instances with 16 agents and 8 values per agent, considering three classes, sparse, medium and dense (p_1=0.2, p_1=0.5 and p_1=0.8). The largest differences between algorithms appear at the complexity peak. Table 4 reports results averaged over 250 executions (50 instances × 5 random seeds) of ABT and ABT_{hyb} without (left) and with delays (right), at the complexity peak.

Table 4. Number of non concurrent constraint checks and messages required by ABT and ABT_{hyb} for random $DisCSP$s, without (left) and with (right) message delays

	ABT		ABT_{hyb}		d-ABT		d-ABT_{hyb}	
p_1	$nccc$	msg	$nccc$	msg	$nccc$	msg	$nccc$	msg
0.20	5,065	6,010	4,595	5,724	4,385	6,375	4,168	5,875
0.50	36,552	33,016	26,558	26,178	29,547	38,097	25,088	30,444
0.80	90,933	63,209	56,566	45,587	68,855	78,557	55,501	58,081

Table 5. Number of messages of ABT and ABT_{hyb} for random $DisCSP$s without message delays

	ABT				ABT_{hyb}			
p_1	ok?	ngd	adl	*obs* ngd (%)	ok?	ngd	adl	*obs* ngd (%)
0.20	4,378	1,606	26	394 (24.53%)	4,298	1,401	26	138 (9.8%)
0.50	25,739	7,238	39	2,442 (33.73%)	21,378	4,762	38	873 (18.33%)
0.80	49,624	13,566	19	5,411 (39.8%)	37,647	7,921	19	1,939 (24.5%)

Regarding results without message delays, ABT_{hyb} is always better than ABT for the three classes, in both computation effort and communication cost. The improvement of ABT_{hyb} over ABT increases when p_1 increases. ABT_{hyb} saves 4.8%, 20.7% and 27.9% of the total number of messages that ABT sends for sparse, medium and dense classes, respectively. This is due to the collective effect already described for the distributed n-queens. This is confirmed by Table 5, that contains the number of messages per type. Waiting states in ABT_{hyb} allow agents to have a more updated information.

Regarding results with message delays, the relative performance remains the same, ABT being less efficient than ABT-hyb. In terms of msg, ABT_{hyb} outperforms ABT by 7.8%, 20.1% and 26.1% for sparse, medium and dense problem classes, respectively. The benefits of adding synchronization points in ABT remains almost the same when considering delays, even when obsolete **ngd** messages increase (Table 6).

5.3   Distributed Sensor-Mobile Problems

The Distributed Sensor-Mobile problem ($SensorDCSP$) is a distributed benchmark based on a real distributed resource allocation problem [4]. It consists of n sensors $\{s_1, s_2, ..., s_n\}$ that track m mobiles $\{v_1, v_2, ...v_m\}$. Each mobile must be tracked by 3 sensors. Each sensor can track at most one mobile. A solution is an assignment of three distinct sensors to each mobile which satisfies visibility and compatibility constraints. The visibility constraint defines the set of mobiles that are visible to each sensor. The compatibility constraint defines the compatibility relationship among sensors.

We encode $SensorDCSP$ in $DisCSP$ as follows. Each agent represents one mobile. Every agent holds exactly one variable. The value domain of each variable is the set

Table 6. Number of messages of ABT and ABT_{hyb} for random $DisCSP$ with message delays

	d-ABT				d-ABT_{hyb}			
p_1	ok?	ngd	adl	*obs* ngd (%)	ok?	ngd	adl	*obs* ngd (%)
0.20	4,486	1,863	26	645 (34.6%)	4,244	1,605	26	344 (21.4%)
0.50	28,580	9,479	39	4,429 (46.7%)	23,690	6,716	38	2,280 (33.9%)
0.80	58,929	19,608	19	10,789 (55.0%)	45,251	12,811	19	5,451 (42.5%)

Table 7. Number of non concurrent constraint checks and messages required by ABT and ABT_{hyb} to solve instances of *SensorDCSP*, without (left) and with (right) message delays

	ABT		ABT_{hyb}		d-ABT		$d\text{-}ABT_{hyb}$	
p_v	nccc	msg	nccc	msg	nccc	msg	nccc	msg
0.20	33,632,099	2,454,366	7,051,764	714,208	32,848,331	2,647,790	7,277,239	799,644
0.50	1,346,669	181,020	1,172,975	144,574	1,306,229	197,149	1,184,653	154,238
0.80	40,744	12,013	31,264	8,961	37,296	13,966	29,919	9,538

Table 8. Number of messages of ABT and ABT_{hyb} solving *SensorDCSP* instances without delays

	ABT			ABT_{hyb}		
p_v	ok?	ngd	obs ngd (%)	ok?	ngd	obs ngd (%)
0.20	1,776,190	678,176	177,012 (26.1%)	550,678	163,530	30,692 (18.7%)
0.50	128,967	52,053	12,661 (24.3%)	104,734	39,840	4,610 (11.6%)
0.80	9,045	2,968	890 (29.9%)	6,899	2,062	366 (16.3%)

Table 9. Number of messages of ABT and ABT_{hyb} solving *SensorDCSP* instances with delays

	d-ABT			$d\text{-}ABT_{hyb}$		
p_v	ok?	ngd	obs ngd (%)	ok?	ngd	obs ngd (%)
0.20	1,861,347	786,443	278,883 (35.5%)	611,747	187,897	57,554 (30.6%)
0.50	137,601	59,548	19,396 (32.6%)	110,975	43,263	7,689 (17.7%)
0.80	10,136	3,829	1,692 (44.2%)	7,276	2,262	597 (26.3%)

all the possible combinations of three sensors that satisfies compatibility and visibility constraints. There is a binary constraint between each pair of variables, which forbids that several mobiles were tracked by the same sensor. *SensorDCSP* instances are generated according to four parameters: the number of sensors (n), the number of mobiles (m), the probability that a mobile is visible for a sensor (p_v), and the probability that two sensors are compatible between them (p_c).

We have evaluated ABT and ABT_{hyb} on *SensorDCSP* instances with 20 sensors, 5 mobiles, three values for p_v (0.20, 0.5 and 0.8) and one value for p_c (corresponding to the most difficult instances). Table 7 presents results averaged over 250 executions (50 instances × 5 random seeds). Again, ABT_{hyb} is always better than ABT in both computation effort and communication cost. Similar results are also observed when considering message delays. Obsolete **ngd** messages are always lower in ABT_{hyb} than in ABT (see Tables 8 and 9).

6   Conclusions

We have presented ABT_{hyb}, a new hybrid algorithm for distributed *CSP* that combines synchronous and asynchronous elements. This algorithm avoids that agents send some redundant messages after backtracking. We have proved that the new algorithm is correct, complete and terminates. There are cases where ABT requires less messages than ABT_{hyb} to solve an instance, but these cases seem to be infrequent in practice, according with the reported experimental results on three benchmarks. For all considered instances, ABT_{hyb} outperforms ABT in terms of the computation effort and communication cost. This improvement is due to the addition of synchronization points when backtracking, which makes ABT_{hyb} less robust than ABT to network failures. But for applications where efficiency is the main concern, ABT_{hyb} seems to be a better candidate than ABT to solve $DisCSP$. Note that ABT_{hyb} benefits remain in presence of message delays.

References

1. Baker, A.B.: The hazards of fancy backtracking. In: Proc. of AAAI 1994, pp. 288–293 (1994)
2. Bessiere, C., Brito, I., Maestre, A., Meseguer, P.: The Asynchronous Backtracking without adding links: a new member in the ABT family. Artifical Intelligence 161, 7–24 (2005)
3. Dechter, R., Pearl, J.: Network-Based Heuristics for Constraint-Satisfaction Problems. Artificial Intelligence 34, 1–38 (1988)
4. Fernández, C., Béjar, R., Krishnamachari, B., Gomes, C.: Communication and Computation in Distributed CSP algorithms. In: Van Hentenryck, P. (ed.) CP 2002. LNCS, vol. 2470, pp. 664–679. Springer, Heidelberg (2002)
5. Meisels, A., Kaplansky, E., Razgon, I., Zivan, R.: Comparing Performance of Distributed Constraint Processing Algorithms. In: AAMAS Workshop on Distributed Constr. Reas., pp. 86–93 (2002)
6. Silaghi, M.C., Sam-Haroud, D., Faltings, B.: Asynchronous Search with Aggregations. In: Proc. of the 17th AAAI, pp. 917–922 (2000)
7. Yokoo, M., Durfee, E., Ishida, T., Kuwabara, K.: Distributed Constraint Satisfaction for Formalizing Distributed Problem Solving. In: Proc. of the 12th. DCS, pp. 614–621 (1992)
8. Yokoo, M.: Asynchronous Weak-Commitment Search for Solving distributed Constraint Satisfaction Problems. In: Montanari, U., Rossi, F. (eds.) CP 1995. LNCS, vol. 976, pp. 88–102. Springer, Heidelberg (1995)
9. Yokoo, M., Durfee, E., Ishida, T., Kuwabara, K.: The Distributed Constraint Satisfaction Problem: Formalization and Algorithms. IEEE Trans. Know. and Data Engin. 10, 673–685 (1998)
10. Yokoo, M., Katsutoshi, H.: Algorithms for Distributed Constraint Satisfaction: A Review. Autonomous Agents and Multi-Agent Systems 3, 185–207 (2000)
11. Zivan, R., Meisels, A.: Synchronous and Asynchronous Search on DisCSPs. In: Proceedings of EUMAS 2003 (2003)

On the Integration of Singleton Consistencies and Look-Ahead Heuristics

Marco Correia and Pedro Barahona

Centro de Inteligncia Artificial, Departamento de Informtica,
Universidade Nova de Lisboa, 2829-516 Caparica, Portugal
{mvc,pb}@di.fct.unl.pt

Abstract. The efficiency of complete solvers depends both on constraint propagation to narrow the domains and some form of complete search. Whereas constraint propagators should achieve a good trade-off between their complexity and the pruning that is obtained, search heuristics take decisions based on information about the state of the problem being solved. In general, these two components are independent and are indeed considered separately. A recent family of algorithms have been proposed to achieve a strong form of consistency called Singleton Consistency (SC). These algorithms perform a limited amount of search and propagation (lookahead) to remove inconsistent values from the variables domains, making SC costly to maintain. This paper follows from the observation that search states being explored while enforcing SC are an important source of information about the future search space which is being ignored. In this paper we discuss the integration of this look-ahead information into variable and value selection heuristics, and show that significant speedups are obtained in a number of standard benchmark problems.

1 Introduction

Complete constraint programming solvers have their efficiency dependent on two complementary components, propagation and search. Constraint propagation is a key component in constraint solving, eliminating values from the domains of the variables with polynomial (local) algorithms. The other component, search, aims at finding solutions in the remaining search space, and is usually driven by heuristics both for selecting the variable to enumerate and the value that is chosen first.

Typically, these components are independent. In particular, heuristics take into account some features of the remaining search space, and some structure of the problem to take decisions. Clearly, the more information there is, the more likely it is to get a good (informed) heuristics. Recently, a lot of attention has been given to a class of algorithms which analyse look-ahead what-if scenarios: what would happen if a variable x takes some value v? Such look-ahead analysis (typically done by subsequently maintaining arc or generalised arc consistency on the constraint network) may detect that value v is not part of any solution,

F. Fages, F. Rossi, and S. Soliman (Eds.): CSCLP 2007, LNAI 5129, pp. 62–75, 2008.
© Springer-Verlag Berlin Heidelberg 2008

and eliminate it from the domain of variable x. This is the purpose of the different variants of Singleton Consistency (SC) [1,4,7,15].

In this paper we propose to go one step further of the above approaches. On the one hand, by recognising that SC propagation is not very cost-effective in general [17], we propose to restrict it to those variables more likely to be chosen by the variable selection heuristics. More specifically, we assume that there are often many variables that can be selected and for which no good criteria exists to discriminate them. This is the case with the first-fail (FF) heuristics, where often there are many variables with 2 values, all connected to the same number of other variables (as is the case with complete graphs). Hence the information gain obtained from SC propagation is used to break the ties between the pre-selected variables.

On the other hand, we attempt to better exploit the information made available by the lookahead procedure, and use it not only to filter values but also to guide search. The idea of exploiting look-ahead information is not new. However in the context of Constraint Programming, look-ahead information has not been fully integrated in subsequent variable and value selection heuristics (see section 5).

In this paper we thus investigate the possibility of integrating Singleton Consistency propagation procedures with look-ahead heuristics, both for variable and value selection heuristics, and analyse the speedups obtained in a number of benchmark problems.

The structure of the paper is the following. In the next section we review some properties of constraint networks. In section 3 we discuss variants of Singleton Consistency, and show how to adapt them to obtain look-ahead information. In section 4 we present a number of benchmark problems and compare the results obtained when using and not using the look-ahead heuristics. In section 5 we report on related work, and finally conclude with a summary of the lessons learned and directions for further research.

2   Notation and Background

A constraint network consists of a set of variables $\mathcal{X}$, a set of domains $\mathcal{D}$, and a set of constraints $\mathcal{C}$. Every variable $x \in \mathcal{X}$ has an associated domain $D(x)$ denoting its possible values. Every k-ary constraint $c \in \mathcal{C}$ is defined over a set of k variables $(x_1, \ldots, x_k)$ by the subset of the Cartesian product $D(x_1) \times \ldots \times D(x_k)$ which are consistent values. The constraint satisfaction problem (CSP) consists in finding an assignment of values to variables such that all constraints are satisfied.

A CSP is arc-consistent iff it has non-empty domains and every consistent instantiation of a variable can be extended to a consistent instantiation involving an additional variable [16]. A problem is generalized arc-consistent (GAC) iff for every value in each variable of a constraint there exist compatible values for all the other variables in the constraint.

Enforcing (generalized) arc consistency is usually not enough for solving a CSP and search must be performed. A large class of search heuristics follow

---
Algorithm 1. SC($\mathcal{X}$,$\mathcal{C}$) : *state*

---
do
   modified ←**false**
   forall $x \in \mathcal{X}$
     modified ← SREVISE(x,$\mathcal{X}$,$\mathcal{C}$) ∨*modified*
     if $D(x) = \emptyset$
       state ←**failed**
       return
     endif
   endfor
while *modified* =**true**
state ←**succeed**

---

the first-fail/best-promise policy (FF/BP) [12], which consists of selecting the variable which more likely leads to a contradiction (FF), and then select the value that has more chances of being part of a solution (BP). For estimating first-failness, heuristics typically select the variable with smaller domain (dom), more constraints attached (deg), more constraints to instantiated variables (bdeg), or combinations (e.g. dom/deg). Best-promise is usually obtained by integrating some knowledge about the structure of the problem.

3 Look-Ahead Pruning Algorithms

3.1 Singleton Consistencies

A CSP P is singleton θ-consistent (SC), iff it has non-empty domains and for any value $a \in dom\,(x)$ of every variable $x \in \mathcal{X}$, the resulting subproblem $P|_{x=a}$ can be made θ-consistent. Most cost-effective singleton consistencies are singleton arc-consistency (SAC) [7] and singleton generalized arc-consistency (SGAC) [17].

To achieve SC in a CSP, procedure SC [7] instantiates each variable to each of its possible values in order to prune those that (after some form of propagation) lead to a domain wipe out (alg. 1).

Once some (inconsistent) value is removed, then there is a chance that a value in a previously revised variable has become inconsistent, and therefore SC must check these variables again. This can happen at most nd times, where n is the number of variables, and d the size of the largest domain, hence SC time complexity is in $O(n^2d^2\Theta)$, Θ being the time complexity of the algorithm that achieves θ-consistency on the constraint network. Variants of this algorithm with the same pruning power but yielding distinct space-time complexity trade-offs have been proposed [1,3,4,15]. A related algorithm considers each variable only once (alg. 2), has better runtime complexity $O(nd\Theta)$, but achieves a weaker consistency, called restricted singleton consistency (RSC) [17].

Note that both algorithms use function SREVISE (alg. 3) which prunes the domain of a single variable by trying all of its possible instantiations.

Algorithm 2. RSC($\mathcal{X}$,$\mathcal{C}$) : $state$

---

forall $x \in \mathcal{X}$
  SREVISE(x,$\mathcal{X}$,$\mathcal{C}$)
  if $D(x) = \emptyset$
    $state \leftarrow$**failed**
    return
  endif
endfor
$state \leftarrow$**succeed**

Algorithm 3. SREVISE(x,$\mathcal{X}$,$\mathcal{C}$) : $modified$

$modified \leftarrow$**false**
forall $a \in D(x)$
  try $x = a$
    $state \leftarrow$PROPAGATE$_\theta$($\mathcal{X}$,$\mathcal{C}$)
  undo $x = a$
  if $state =$**failed**
    $D(x) \leftarrow D(x) \setminus a$
    $modified \leftarrow$**true**
  endif
endfor

3.2 Pruning Decisions

Another possible trade-off between run-time complexity and pruning power is to enforce singleton consistency on a subset of variables $\mathcal{S} \subset \mathcal{X}$. We identified two possible goals which condition the selection of $\mathcal{S}$: filtering and decision making. From a filtering perspective, $\mathcal{S}$ should be the smallest subset where (restricted) singleton consistency can actually prune values, and although this is not known *a priori,* approximations are possible by exploring incrementality and value support [1,4]. On the other hand, $\mathcal{S}$ may be selected for improving the decision making process, in particular of variable selection heuristics that are based on the cardinality of the current domains. In this case, the pruning resulting from enforcing singleton consistency is used as a mechanism to break ties both in the selection of variable and in the choice of the value to enumerate.

Observing the general preference for variable heuristics which select smallest domains first, we propose defining $\mathcal{S}$ as the set of variables whose domain cardinality is below a given threshold d. We denote by RSC$_d$($\mathcal{X}$, $\mathcal{C}$) and SC$_d$($\mathcal{X}$, $\mathcal{C}$), respectively, the algorithms RSC($\mathcal{X}_{|D| \leq d}$, $\mathcal{C}$) and SC($\mathcal{X}_{|D| \leq d}$, $\mathcal{C}$), where $\mathcal{X}_{|D| \leq d}$ is the subset of variables in $\mathcal{X}$ having domains with cardinality less or equal to d.

A further step in integrating singleton consistencies with search heuristics is to explore information regarding the subproblems that are generated each time a value is tested for consistency. We propose a class of look-ahead heuristics (LA) for any CSP P which reason over the size of its solution space, given by a function $\sigma(P)$, collected while enforcing singleton consistency. Although

Algorithm 4. sREVISEINFO(x,$\mathcal{X}$,$\mathcal{C}$,$info$) : $modified$

$modified \leftarrow$ **false**
forall $a \in D(x)$
   try $x = a$
      $state \leftarrow$ PROPAGATE$_\theta$($\mathcal{X}$,$\mathcal{C}$)
      $info[x,a] \leftarrow$ COLLECTINFO($\mathcal{X}$,$\mathcal{C}$)
   undo $x = a$
   if $state =$**failed**
      $D(x) \leftarrow D(x) \setminus a$
      $modified \leftarrow$**true**
   endif
endfor

Algorithm 5. SEARCH($\mathcal{X}$,$\mathcal{C}$) : $state$

$info \leftarrow \emptyset$
if SC($\mathcal{X},\mathcal{C}$) =**fail**
   $state \leftarrow$**fail**
   return
endif
if $\forall_x : |D(x)| = 1$
   $state \leftarrow$**succeed**
   return
endif
$x \leftarrow$ SELECTVARIABLE($\mathcal{X}$,$info$)
$a \leftarrow$ SELECTVALUE(x,$info$)
$state \leftarrow$ SEARCH($\mathcal{X}$,$\mathcal{C} \cup (x = a)$) **or** SEARCH($\mathcal{X}$,$\mathcal{C} \cup (x \neq a)$)

there is no known polynomial algorithm for computing σ (finding the number of solutions of a CSP is a #P-complete problem), there exists a number of naive as well as more sophisticated approximation functions [10,13]. We conjecture that by estimating the size of the solution space for each possible instantiation, i.e. $\sigma(P|_{x=a})$, there is an opportunity for making more informed decisions that will exhibit both better first-failness and best-promise behaviour. Moreover, the class of approximations of σ presented below are easy to compute, do not add complexity to the cost of generating the subproblems, and only requires a slight modification of the sREVISE algorithm.

The sREVISEINFO algorithm (alg. 4) stores in a table ($info$) relevant information to the specific subproblem being considered in each loop iteration. In our case, $info$ is an estimation of the subproblem solution space, more formally $info[x,a] = \sigma'(P|_{x=a})$ where $\sigma' \approx \sigma$. The table is initialized before singleton consistency enforcement, computed after propagation, and handed to the SELECTVARIABLE and SELECTVALUE functions as shown in algorithm 5.

There are several possible definitions for these functions associated with how they integrate the collected information. Regarding the selection of variable for a given CSP P, we identified two FF heuristics which are cheap and easy to compute:

$$var_1 (P) = \arg \min_{x \in \mathcal{X}(P)} \left(\sum_{a \in D(x)} \sigma' \left(P|_{x=a} \right) \right)$$

$$var_2 (P) = \arg \min_{x \in \mathcal{X}(P)} \left(\max_{a \in D(x)} \sigma' \left(P|_{x=a} \right) \right)$$

Informally, var_1 gives preference for the variable with a smaller sum of the number of solutions for every possible instantiation, while var_2 selects the variable whose instantiation with maximum number of solutions is the minimum among all variables. For the selection of value for some variable x, a possible BP heuristic is

$$val_1 (P, x) = \arg \max_{a \in D(x)} \left(\sigma' \left(P|_{x=a} \right) \right)$$

which simply prefers the instantiation that prunes less solutions from the remaining search space.

Functions var_1 and val_1 correspond to the minimize promise variable heuristic and maximize promise value heuristic defined in [9]. Please note that we do not claim these are the best options for the estimation of the search space or the number of solutions. We have simply adopted them for simplicity and for testing the concept (more discussion on section 6).

4 Experimental Results

A theoretical analysis on the adequacy of these heuristics as FF or BP candidates is needed, but hard to accomplish. Alternatively, in this section we attempt to give some empirical evidence of the quality of these heuristics by presenting the results of using them combined with constraint propagation and backtracking search (BT) on a set of typical CSP benchmarks.

The set of heuristics selected for comparison was chosen in order to provide some insight on the adequacy of enforcing SC on a subset of variables as a good trade-off between propagation and search and on the impact of integrating LA information in the variable and value selection heuristics. As a side effect, we tried to confirm previous results on the classes of instances where SC is cost effective and on the performance of RSC regarding SC.

As a first attempt at measuring the potential of LA heuristics, a simple measure was used for estimating the number of solutions in a given subproblem:

$$\sigma' = \sum_{x \in \mathcal{X}} \log_2 \left(D(x) \right)$$

which informally expresses that the number of solutions is correlated to the size of the subproblem search space[1]. Although this is a very rough estimate, we are assuming that it could nevertheless provide valuable information to compare alternatives (see section 6).

---

[1] We use the logarithm since the size of search space can be a very large number.

As a baseline for comparison we used the dom variable selection heuristic (see section 2) without any kind of singleton consistency enforcing. The other elements of the test set are the possible combinations of enforcing SC, RSC, SC2 or RSC2 with the dom or LA heuristics. The SC2 and RSC2 tests implement the $SC_d(\mathcal{X}, \mathcal{C})$ and $RSC_d(\mathcal{X}, \mathcal{C})$ strategies with $d = 2$, the threshold for which most interesting results were obtained. The LA heuristics implement the proposed functions var_1 and val_1. Each combination is thus denoted by $a+b+c$, where a states the type of singleton consistency enforced (or is absent if none), b specifies the variable heuristic and c the value heuristic. For example, sc+dom+min performs singleton consistency and then instantiates the variable with the smallest domain to the minimum value in its domain.

In the following experiments all times are given in seconds, and represent the time needed for finding the first solution. The column 'ratio', when present, refers to the average CPU time of the current heuristic over the baseline, which is always the CPU time of the dom heuristic. Data presented in the following charts was interpolated using a Bzier smoothing curve.

Tests regarding sections 4.1 and 4.2 were performed on a Pentium4, 3.4GHz with 1Gb RAM, while the results presented in section 4.3 were obtained on a Pentium4, 1.7GHz with 512Mb RAM.

4.1   Graph Coloring

Graph coloring consists of trying to assign n colors to m nodes of a given graph such that no pair of connected nodes have the same color. In this section we evaluate the performance of the presented heuristics in two sets of 100 instances of 10-colorable graphs, respectively with 50 and 55 nodes, generated using Joseph Culberson's k-colorable graph generator [6].

A CSP for solving the graph coloring problem was modelled by using variables to represent each node and values to define its color. Difference binary constraints were posted for every pair of connected nodes.

The average degree of a node in the graph d, i.e. the probability that each node is connected to every other node, has been used for describing the phase transition in graph coloring problems [5]. In this experiment we started by determining empirically the phase transition to be near $d = 0.6$, and then generated 100 random instances varying d uniformly in the range $[0.5 \ldots 0.7]$.

Figure 1 compares the search effort using each heuristic on the smallest graph problem, with a timeout of 300 seconds. These results clearly divide the heuristics into two sets, the set where SC and RSC was used being much better than the other on the hard instances. Since the ranking within the best set was not so clear, a second experiment on the larger and more difficult problem was performed using only these four heuristics, with a larger timeout of 900 seconds. The results of these tests are shown graphically on fig. 2, and also given in detail in table 1.

This second set of experiments shows that RSC+LA is better on the most difficult instances (almost by an order of magnitude), while the others have quite similar efficiency.

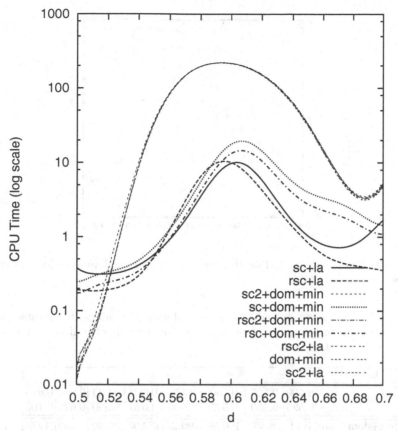

Fig. 1. CPU time spent in finding the first solution of random 10-colorable graph instances with size 50

4.2   Random CSPs

Randomly generated CSPs have been widely used experimentally, for instance to compare different solution algorithms. In this section we evaluate the look-ahead heuristics on several random n-ary CSPs. These problems were generated using model C [11] generalized to n-ary CSPs, that is, each instance is defined by a 5-tuple $\langle n, d, a, p_1, p_2 \rangle$, where n is the number of variables, d is the uniform size of the domains, a is the uniform constraint arity, p_1 is the density of the constraint graph, and p_2 the looseness of the constraints.

These tests evaluate the performance of the several heuristics in a set of random instances near the phase transition. For this task we used the constrainedness measure κ [10] for the case where all constraints have the same looseness and all domains have the same size:

$$\kappa = \frac{- |\mathcal{C}| \log_2 (p_2)}{n \log_2 d}$$

where $|\mathcal{C}|$ is the number of n-ary constraints.

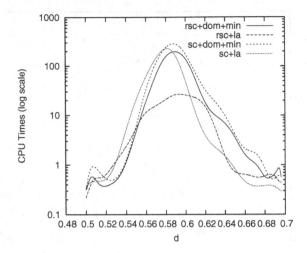

Fig. 2. CPU time spent in finding the first solution of random 10-colorable graph instances with size 55

Table 1. CPU time spent in finding the first solution of random 10-colorable graph instances with size 55. Columns show averages for intervals of uniform variation of constraint tightness d.

heuristic	d							
	0.500 0.525	0.525 0.550	0.550 0.575	0.575 0.600	0.600 0.625	0.625 0.650	0.650 0.675	0.675 0.700
rsc+dom+min	0.71	8.46	160.59	499.27	179.93	26.60	13.59	1.00
sc+dom+min	0.93	**0.83**	183.58	624.11	184.67	46.13	21.43	0.47
rsc+la	1.18	5.03	**148.92**	**67.13**	**72.62**	**26.44**	0.73	0.72
sc+la	**0.46**	104.29	320.35	603.55	107.27	69.51	**0.36**	**0.37**

We started by fixing n, d and a arbitrarily to 50, 5 and 3 respectively, and then computed 100 values for p_2 uniformly in the range $[0.1 \ldots 0.8]$. For each of these values, a value of p_1 was used such that $\kappa = 0.95$ (problems in the phase transition have typically $\kappa \approx 1$). The value of p_1, given by

$$p_1 = -\kappa \frac{n \log_2 d}{\log_2 p_2} \times \frac{a!\,(n-a)!}{n!}$$

is computed from the first formula and by noting that p_1 is the fraction of constraints over all possible constraints in the constraint graph, i.e.

$$p_1 = |\mathcal{C}|\, \frac{a!\,(n-a)!}{n!}$$

Solutions were stored as positive table constraints and GAC-Schema [2] was used for filtering. The timeout was set to 600 seconds.

Table 2. CPU time spent in finding the first solution of random CSP instances Columns show averages for intervals of uniform variation of constraint looseness p_2

heuristic	p_2						
	0.1-0.2	0.2-0.3	0.3-0.4	0.4-0.5	0.5-0.6	0.6-0.7	0.7-0.8
dom+min	**0.06**	**0.34**	2.98	12.86	52.84	236.82	377.18
rsc2+dom+min	0.10	0.58	4.69	18.79	73.61	279.86	429.22
rsc+dom+min	0.21	1.09	5.35	25.03	97.32	319.71	471.83
sc2+dom+min	0.11	0.64	4.95	20.45	79.78	289.59	438.92
sc+dom+min	0.27	1.28	6.27	29.28	112.15	341.82	492.76
rsc2+la	**0.09**	0.45	3.47	11.85	53.64	237.19	373.80
rsc+la	0.11	**0.37**	**1.43**	**3.28**	**21.86**	**71.10**	**99.93**
sc2+la	0.10	0.50	3.78	12.92	58.60	249.03	388.77
sc+la	0.15	0.47	**1.75**	**4.04**	**26.06**	**82.60**	**115.13**

Table 2 shows the results obtained. In figure 3 the performances of the most interesting heuristics are presented graphically.

Besides the rsc+dom+min and sc+dom+min heuristics which always performed worse than the others, there seems to be a change of ranking around $p_2 \approx 0.4$, with the dom+min dominating on the dense instances, and LA heuristics 3-4 times faster on the sparse zone. RSC+LA and SC+LA are consistently close across all instances, being RSC slightly better.

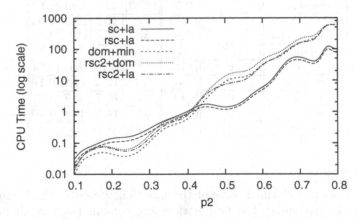

Fig. 3. CPU time spent in finding the first solution of random CSP instances

4.3 Partial Latin Squares

Latin squares is a well known benchmark which combines randomness and structure [19]. The problem consists in placing the elements $1 \ldots N$ in a $N \times N$ grid, such that each element occurs exactly once on the same row or column. A partial Latin squares (or quasigroup completion) problem is a Latin squares problem with a number of preassigned cells, and the goal is to complete the puzzle.

The problem was modelled using the direct encoding, i.e. using an all-different (GAC) constraint for every row and column. The dual encoding model, as proposed in [8], was also considered but never improved over the direct model on the presented instances. The value selection heuristic used in conjunction with the dom variable selection heuristic, denoted as mc (minimum-conflicts), selects the value which occurs less in the same row and column of the variable to instantiate. This is reported to be the best known value selection heuristic for this problem [8].

We generated 200 instances of a satisfiable partial Latin squares of size 30, with 312 cells preassigned, using lsencode-v1.1 [14], a widely used random quasigroup completion problem generator. The timeout was set to 900 seconds.

Results are presented on table 3. In this problem there is a clear evidence of the rsc2+la and sc2+la heuristics compared to every other. Besides the fact that they are over 5 times faster than the other alternatives, they are also the most robust, as shown by their lower standard deviations as well as the absence of time out instances.

Table 3. Running times and number of fails for the pls-30-312 problem. Last column shows the ratio between the average time of each heuristic over the average time of the baseline, which is the dom+mc heuristic.

heuristic	#timeouts	#fails		time		
		avg	std	avg	std	ratio
dom+mc	5	18658	43583	66.7	169	1
rsc2+dom+mc	5	235	436	70.2	156.4	1.05
rsc+dom+mc	5	24	49	89.3	159.2	1.34
sc2+dom+mc	10	174	330	100.5	207.1	1.51
sc+dom+mc	6	**15**	**28**	122.8	180.3	1.84
rsc2+la	**0**	51	127	**12.2**	**19.9**	**0.18**
rsc+la	**0**	14	35	67.7	47.7	1.02
sc2+la	**0**	43	109	**15.6**	**26.4**	**0.23**
sc+la	4	**11**	**29**	104.6	134.4	1.57

4.4   Discussion

The results obtained clarified some of the questions posed in the beginning of this section. In particular, the best performing combinations in all problems were always obtained using LA information, so this approach has clearly some potential to be explored more thoroughly.

Regarding the use of SC on a subset of variables, the results so far are not conclusive. Heuristics that restrict SC maintenance to only 2 valued variables performed badly both on the graph coloring and random problems, but clearly outperformed all others on the Latin squares problem. We think that this behaviour may be connected with the number of times these heuristics have a chance to break ties both in the selection of variable and value. The cardinality of the domains should have impact on the number of decisions having the same preference for FF heuristics, in particular the dom heuristic. The average number of values in the Latin squares problem is very low (around 3) since most variables are instantiated, so these heuristics would have more chance to make a difference here than on the other problems which have larger cardinality (5 and 10). The same argument may apply to the value selection heuristic if we note that the selection of value is more important in problems with some structure, which would again favour the Latin squares problem.

The remaining aspects of the results obtained are in accordance with the extensive analysis of singleton consistencies described in [17]. On the question of cost-effectiveness of RSC we obtained similar positive results, in fact it was slightly better than SC on all instances. Its combination with LA was the most successful, outperforming the others in the hard instances of every problem.

In the class of random problems, their work concludes that singleton consistencies are only useful in the sparse instances. Our results also confirm this. Generally, the claim that SC can be very expensive to maintain seems true in our experiments except when using combined with LA heuristics. This provides some evidence that the good behaviour of SC+LA observed relies more strongly on correct decisions rather than on the filtering achieved.

5   Related Work

The work of [18] suggests improving the variable selection heuristic based on the impact each variable assignment had on past search states. The impact is defined as the ratio of search size reduction achieved when propagating the assignment. In their paper the use of a specific look-ahead procedure for measuring this impact is regarded as costly, and depreciated in favour of a method that accumulates this information across distinct search branches and/or search iterations (restarts). Their results show that the method eventually converges to a good variable ordering (the value selection heuristic is not considered).

In [13], belief updating techniques are used to estimate the likelihood of a value belonging to some solution. These likelihoods are then used to improve the value selection heuristic and as propagation: if it decreases to zero, the value is discarded from the domain. However, the integration of this kind of propagation with common local propagation algorithms is not explored in that paper.

6   Conclusion

In this paper we presented an approach that incorporates look ahead information for directing backtracking search, and suggested that this could be largely done at no extra cost by taking advantage of the work already performed by singleton consistency enforcing algorithms. We described how such a framework could extend existing SC and RSC algorithms by requiring only minimal modifications. Additionally, a less expensive form of SC was revisited, and a new one proposed which involves revising only a subset of variables. Empirical tests with two common benchmarks and with randomly generated CSPs showed promising results on instances near the phase transition. Finally, results were analysed and matched against those previously obtained by other researchers.

There are a number of open questions and future work directions. As discussed in the previous section, tests which use singleton consistency on a subset of variables defined by its cardinality were not consistently better or worse than the others, but may be very beneficial sometimes. We think this deserves more investigation, namely testing with more structured problems, using a distinct selection criteria (other than domain cardinality), and selective performing singleton consistency less often by reusing previously computed information (in the *info* table).

The most promising direction for future work is improving the FF and BP measures. Look-ahead heuristics presented above use rather naive estimation of number of solutions for a given subproblem compared to, for example, the κ measure introduced in [10], or the probabilistic inference methods described in [13]. The κ measure, for example, takes into account the individual tightness of each constraint and the global density of the constraint graph. Their work shows strong evidence for best performance of this measure compared with standard FF heuristics, but also point out that the complexity of its computation may lead to suboptimal results in general CSP solving (the results reported are when using forward-checking). Given that we perform a stronger form of propagation and have look-ahead information available, the cost for computing κ may be worth while. We intend to investigate this in the future.

Other improvements include the use of faster singleton consistency enforcing algorithms [1,4], which should be orthogonal to the results presented here, and the use of constructive disjunction during the maintenance of SC, by pruning values from the domains of a variable that does not appear in the state of the problem for all values of another variable.

We think the results obtained so far are quite promising and justify further research along the outlined directions.

References

1. Barták, R., Erben, R.: A new algorithm for singleton arc consistency. In: Barr, V., Markov, Z. (eds.) Proceedings of the Seventeenth International Florida Artificial Intelligence Research Society Conference (FLAIRS 2004), Miami Beach, Florida, USA. AAAI Press, Menlo Park (2004)
2. Bessiére, C., Régin, J.-C.: Arc consistency for general constraint networks: preliminary results. In: Proceedings of IJCAI 1997, Nagoya, Japan, pp. 398–404 (1997)
3. Bessière, C., Debruyne, R.: Theoretical analysis of singleton arc consistency. In: Proceedings of ECAI 2004 (2004)
4. Bessière, C., Debruyne, R.: Optimal and suboptimal singleton arc consistency algorithms. In: Proceedings of IJCAI 2005 (2005)
5. Cheeseman, P., Kanefsky, B., Taylor, W.M.: Where the really hard problems are. In: Proceeding of IJCAI 1991, pp. 331–340 (1991)
6. Culberson, J.: Graph coloring resources. on-line, http://web.cs.ualberta.ca/joe/Coloring/Generators/generate.html

7. Debruyne, R., Bessière, C.: Some practicable filtering techniques for the constraint satisfaction problem. In: Proceedings of IJCAI 1997, pp. 412–417 (1997)
8. del Val, D., Cebrian: Redundant modeling for the quasigroup completion problem. In: Rossi, F. (ed.) CP 2003. LNCS, vol. 2833, Springer, Heidelberg (2003)
9. Geelen, P.A.: Dual viewpoint heuristics for binary constraint satisfaction problems. In: Proceedings of ECAI 1992, pp. 31–35. John Wiley & Sons, Inc., New York (1992)
10. Gent, I.P., MacIntyre, E., Prosser, P., Walsh, T.: The constrainedness of search. In: Proceedings of AAAI 1996, vol. 1, pp. 246–252 (1996)
11. Gent, I.P., MacIntyre, E., Prosser, P., Smith, B.M., Walsh, T.: Random constraint satisfaction: Flaws and structure. Constraints 6(4), 345–372 (2001)
12. Haralick, R.M., Elliott, G.L.: Increasing tree search efficiency for constraint satisfaction problems. Artificial Intelligence 14, 263–313 (1980)
13. Kask, K., Dechter, R., Gogate, V.: New look-ahead schemes for constraint satisfaction. In: Proceeding of AMAI 2004 (2004)
14. Kautz, Ruan, Achlioptas, Gomes, Selman, Stickel.: Balance and filtering in structured satisfiable problems. In: Proceedings of IJCAI 2001 (2001)
15. Lecoutre, C., Cardon, S.: A greedy approach to establish singleton arc consistency. In: Kaelbling, L.P., Saffiotti, A. (eds.) Proceedings of IJCAI 2005, pp. 199–204. Professional Book Center (2005)
16. Mackworth, A.K., Freuder, E.C.: The complexity of some polynomial network consistency algorithms for constraint satisfaction problems. Artificial Intelligence 25, 65–74 (1985)
17. Prosser, P., Stergiou, K., Walsh, T.: Singleton consistencies. In: Dechter, R. (ed.) CP 2000. LNCS, vol. 1894, pp. 353–368. Springer, Heidelberg (2000)
18. Refalo, P.: Impact-based search strategies for constraint programming. In: Wallace, M. (ed.) CP 2004. LNCS, vol. 3258, pp. 557–571. Springer, Heidelberg (2004)
19. Shaw, P., Stergiou, K., Walsh, T.: Arc consistency and quasigroup completion. In: Proceedings of ECAI 1998 workshop on non-binary constraints, March 14 (1998)

Combining Two Structured Domains for Modeling Various Graph Matching Problems

Yves Deville[1], Grégoire Dooms[2], and Stéphane Zampelli[1]

[1] Department of Computing Science and Engineering, Université catholique de Louvain,
B-1348 Louvain-la-Neuve - Belgium
{Yves.Deville,Stephane.Zampelli}@uclouvain.be
[2] Department of Computer Science, Brown University, Box 1910, Providence, RI 02912, USA
gdooms@cs.brown.edu

Abstract. Graph pattern matching is a central application in many fields. In various areas, the structure of the pattern can only be approximated and exact matching is then too accurate. We focus here on approximations declared by the user within the pattern (optional nodes and forbidden arcs), covering graph/subgraph mono/isomorphism problems. In this paper, we show how the integration of two domains of computation over countable structures, *graphs* and *maps*, can be used for modeling and solving various graph matching problems from the simple graph isomorphism to approximate graph matching. To achieve this, we extend map variables allowing the domain and range to be non-fixed and constrained. We describe how such extended maps are designed then realized on top of finite domain and finite set variables with specific propagators. We show how a single monomorphism constraint is sufficient to model and solve those multiples graph matching problems. Furthermore, our experimental results show that our CP approach is competitive with a state of the art algorithm for subgraph isomorphism.

1 Introduction

Graph pattern matching is a central application in many fields [1]. Many different types of algorithms have been proposed, ranging from general methods to specific algorithms for particular types of graphs. In constraint programming, several authors [2,3] have shown that graph matching can be formulated as a CSP problem, and argued that constraint programming could be a powerful tool to handle its combinatorial complexity.

In many areas, the structure of the pattern can only be approximated and exact matching is then far too stringent. Approximate matching is a possible solution, and can be handled in several ways. In a first approach, the matching algorithm may allow part of the pattern to mismatch the target graph (e.g. [4,5,6]). The matching problem can then be stated in a probabilistic framework (see, e.g. [7]). In a second approach, the approximations are declared by the user within the pattern, stating which part could be discarded (see, e.g. [8,9]). This approach is especially useful in fields, such as bioinformatics, where one faces a mixture of precise and imprecise knowledge of the pattern structures. In this approach, which will be followed in this paper, the user is able to choose parts of the pattern open to approximation.

Within the CSP framework, a model for graph isomorphism has been proposed by Sorlin et al. [10], and by Rudolf [3] and Valiente et al. [2] for graph monomorphism.

F. Fages, F. Rossi, and S. Soliman (Eds.): CSCLP 2007, LNAI 5129, pp. 76–90, 2008.
© Springer-Verlag Berlin Heidelberg 2008

Subgraph isomorphism in the context of the SBDD method for symmetry breaking is shortly described in [11]. We also proposed in [9] a CSP model for approximate graph matching, but without graph and map variables. Our propagators for monomorphism are based on these works. A declarative view of matching has also been proposed in [12] in the context of XML queries.

In constraint programming, two domains of computation over countable structures have received recent attention : graphs and maps. In CP(Graph) [13], graph variables, and constraints on these variables are described (see also [14,15] for similar ideas). CP(Graph) can be used to express and solve combinatorial graph problems modeled as constrained subgraph extraction problems. In [16,17], function variables are proposed, but the domain and range are limited to ground sets. Such function variables are useful for modeling problems such as warehouse location.

In this paper, we propose an extension to function variables by generalizing them to non-fixed range and domain (source and target set). We call this extension CP(Map) and show how approximate graph matching can be modeled and solved, within the CSP framework, on top of CP(Graph+Map).

Contributions. The main contributions of this work are the following:

- Extension of function variables, where the domain and range of the mapping are not limited to ground sets, but can be finite set variables. Introduction of the *MapVar* and *Map* constraints which allow to use the non-fixed feature of our map variables.
- Demonstration of how a single constraint is able to express a wide range of graph matching problems thanks to three high-level structured variables. In particular, we show how switching a parameter from a fixed graph to a graph interval opens a new spectrum of matching problems. We show how additional constraints imposed on this graph interval enable the expression of hybrid problems such as approximate graph matching. The beauty and originality of this approach resides in that those problems are either new or were always treated separately, illustrating the expressive power and generality of constraint programming.
- Experimental evaluation of our CP approach. We show that this modeling exercise is not only aesthetic but is actually competitive with the current state of the art in subgraph isomorphism (vflib). The genericity of the approach does not hinder the efficiency of the solver. On a standard benchmark set, we show that our approach solves in a given time limit a fourth of the instances which cannot be solved by vflib while only spending between 9% and 22% more time on instances solved by the two competing approaches.

The next section describes the basic idea behind the CP(Graph) framework. CP(Map), our extension to function variables in CP is described in Section 3. Approximate graph matching is defined in Section 4, and its modeling within CP(Graph+Map) is handled in Section 5. Section 6 analyses experimental results, and Section 7 concludes this paper.

2 CP(Graph)

Graphs have been around since the first years of constraint programming. Some problems involving undetermined graphs have been formulated using either binary variables, sets ([14,15]) or integers (successor variables e.g. in [18,19]). CP(Graph) [13] unifies those models by recognizing a common structure: Graph variables are variables whose domain ranges over a set of graphs and as with set variables [20,16], this set of graphs is represented by a graph interval $[\underline{D}(G), \overline{D}(G)]$ where $\underline{D}(G)$, the greatest lower bound (glb) and $\overline{D}(G)$, the least upper bound (lub) are two graphs with $\underline{D}(G)$ a subgraph of $\overline{D}(G)$ (we write $\underline{D}(G) \subseteq \overline{D}(G)$). These two bounds are referred to as the lower and the upper bound. The lower bound $\underline{D}(G)$ is the set of all nodes and arcs which *must* be part of the graph in a solution while the upper bound $\overline{D}(G)$ is the set of all nodes and arcs which could be part of the graph in some solution. The domain of a graph variable with $D(G) = [\underline{D}(G), \overline{D}(G)]$ is the set of graphs g with $\underline{D}(G) \subseteq g \subseteq \overline{D}(G)$. Here, g is used to denote a constant graph and G is used to denote a graph variable. This notation is used throughout this paper: in CSP, lowercase letters denote constants and uppercase letters denote domain variables.

Graph variables can be implemented using a dedicated data-structure or translated into set variables, integer variables or binary variables. For instance, a graph variable G can be modeled as a set of nodes N and a set of arcs E with an additional constraint enforcing the relation $E \subseteq N \times N$. Whatever the graph variable implementation, two basic constraints $Nodes(G, SN)$ and $Arcs(G, SA)$ allow to access respectively the set of nodes and the set of arcs of the graph variable. To simplify the notation the expression $Nodes(G)$ is used to represent a set variable constrained to be equal to the set of nodes of G. A similar notation is used for arcs.

Various constraints have been defined over such graph variables (or their preceding specialized models); see for instance the cycle [18], tree [21], path [22,23], minimum spanning tree [24] or spanning tree optimization constraint [25]. In the remainder of this article, we only use the two simple constraints $Subgraph(G_1, G_2)$ (also denoted $G_1 \subseteq G_2$) and $InducedSubgraph(G_1, G_2)$ (also denoted $G_1 \subseteq^* G_2$). $G_1 \subseteq G_2$ holds if G_1 is a subgraph of G_2, its propagator enforces that the lower and upper bounds of G_1 are subgraphs of the lower bound and upper bounds of G_2 respectively. The constraint $G_1 \subseteq^* G_2$ states that G_1 is the node-induced subgraph of G_2. It holds if G_1 is a subgraph of G_2 such that for each arc a of G_2 whose end-nodes are in G_1, a is also in G_1.

3   CP(Map)

The value of a map variable is a mapping from a domain set to a range set. The domain of a map variable is thus a set of mappings. Map variables were first introduced in CP in [16] where Gervet defines relation variables. However, the domain and the range of the relations were limited to ground finite sets. Map variables were also introduced as high level type constructors, simplifying the modeling of combinatorial optimization problems. This was first defined in [17] as a relation or map variable M from set v into a set w, where supersets of v and w must be known. Such map variables are then compiled into OPL. This idea is developed in [26], but the domain and range of a map variable are limited to ground sets. Relation and map variables are also described in [27] as a useful abstraction in constraint modeling. Rules are proposed for refining constraints on these complex variables into constraints on finite domain and finite set variables. Map variables were also introduced in modeling languages such as ALICE

[28], REFINE [29] and NP-SPEC [30]. To the best of our knowledge, map variables were not yet introduced directly in a CP language. One challenge is then to extend current CP languages to allow map variables as well as constraints on these variables.

In the remaining of this section, we show how a CP(Map) extension can be realized on top of finite domain and finite set variables.

3.1 The Map Domain

We consider the domain of total surjective functions. Given two elements $m_1 : s_1 \to t_1$ and $m_2 : s_2 \to t_2$, where s_1, s_2, t_1, t_2 are sets, we have $m_1 \subseteq m_2$ iff $s_1 \subseteq s_2 \wedge t_1 \subseteq t_2 \wedge \forall x \in s_1 : m_1(x) = m_2(x)$. We also have that $m = \mathrm{glb}(m_1, m_2)$ is a map $m : s \to t$ with $s = \{x \in s_1 \cap s_2 \mid m_1(x) = m_2(x)\}$, $t = \{v \mid \exists x \in s : m_1(x) = v\}$, and $\forall x \in s : m(x) = m_1(x) = m_2(x)$. The lub between two elements m_1, m_2 exists only if $\forall x \in s_1 \cap s_2 : m_1(x) = m_2(x)$. In that case the lub is a map $m : s \to t$ with $m(x) = m_1(x)$ if $x \in s_1$, and $m(x) = m_2(x)$ if $x \in s_2$, $s = s_1 \cup s_2$, and $t = \{v \mid \exists x \in s : m(x) = v\}$. The domain of total surjective functions is then a meet semi lattice, that is a semi lattice where every pairs of elements has a glb.

3.2 Map Variables and the MapVar Constraint

A map variable is declared with the constraint $MapVar(M, S, T)$, where M is the map variable and S, T are finite set variables of Cardinal [31]. The domain of M is all the *total surjective* functions from s to t, where s, t are in the domain of S, T. We call S the *source set* of M, and T the *target set* of M. When M is instantiated (when its domain is a singleton), the source set and the target set of M are ground sets corresponding to the domain and the range of the mapping. As usual, the domain of a set variable S is represented by a set interval $[\underline{D}(S), \overline{D}(S)]$, the set of sets s with $\underline{D}(S) \subseteq s \subseteq \overline{D}(S)$.

Example Let M be a map variable declared in $MapVar(M, S, T)$, with $dom(S) = [\{8\}, \{4, 6, 8\}]$ and $dom(T) = [\{\}, \{1, 2, 4\}]$. A possible instance of M is $\{4 \to 1, 8 \to 4\}$. On this instance, $S = \{4, 8\}$, and $T = \{1, 4\}$. Another instance is $M = \{4 \to 1, 8 \to 1\}$, $S = \{4, 8\}$, and $T = \{1\}$.

Map variables can be used for defining various kinds of mappings, such as :

– Surjective function : $SurjectFct(M, S, T) \equiv MapVar(M, S, T)$.
– Bijective function : $BijectFct(M, S, T) \equiv SurjectFct(M, S, T)$
  $\wedge \forall i, j \in S : i \neq j \Rightarrow M(i) \neq M(j)$.
– Injective function : $InjectFct(M, S, T) \equiv T' \subseteq T \wedge BijectFct(M, S, T')$
– Total function : $TotalFct(M, S, T) \equiv T' \subseteq T \wedge SurjectFct(M, S, T')$
– Partial function : $PartialFct(M, S, T) \equiv S' \subseteq S \wedge TotalFct(M, S', T)$

In order to access individual elements of the map, we define the constraint $Map(M, X, V)$, where X and V are finite domain variables. Given a map variable declared with $MapVar(M, S, T)$, the constraint $Map(M, X, V)$ holds when $X \in S \wedge V \in T \wedge M(X) = V$. We also define the constraint $M1 \subseteq M2$.

3.3 Implementing Map Variables in a Finite Domain Solver

When a map variable M is declared by $MapVar(M, S, T)$, a finite domain (FD) variable M_x is associated to each element x of the upper bound of the source set ($\overline{D}(S)$).

The semantics of these FD variables is simple : M_x represents $M(x)$, the image of x through the function M. Since the source set S can be non-fixed, x might eventually not be in S and its image would not be defined. A special value $\bot$ is used for this purpose. The relationship between the domain of each variable M_x and the set variables S and T can be stated as follows :

- (1) $S = \{x \mid M_x \neq \bot\}$ (M is total)
- (2) $T = \{v \mid \exists x : M_x = v \neq \bot\}$ (M is surjective)

Given $MapVar(M, S, T)$, the domain of M is the set of total surjective functions $m : s \to t$ with $s \in D(S)$, $t \in D(T)$, $\forall x \in s : m(x) \in D(M_x)$, and $\forall x \notin s : \bot \in D(M_x)$.

As can be seen on Figure 1, these variables are stored in an array and accessed by value x through a dictionary data structure (e.g. hashmap) $index$ used to store the index in the array of each value of $\overline{D}(S)$. The initial domain of each FD variable is $\overline{D}(T) \cup \{\bot\}$.

3.4 Additional Constraints and Propagators

Given two map constraints $MapVar(M1, S1, T1)$ and $MapVar(M2, S2, T2)$ the constraint $M1 \subseteq M2$ is implemented as $S1 \subseteq S2 \wedge T1 \subseteq T2 \wedge \forall x \in S1 : M1_x = M2_x$. The last conjunct can be implemented as a set of propagation rules :

- $x \in \underline{D}(S1) \to M1_x = M2_x$
- for each $x \in \overline{D}(S1) \setminus \underline{D}(S1) : M1_x \neq M2_x \to x \notin S1$.

The constraint $Map(M, X, V)$ is translated to $Element(index(X), I, V) \wedge X \in S \wedge V \in T$, where S and T are the source and target sets of M, I is the array representing the FD variables M_x, and $index(X)$ is a finite domain obtained by taking the index of each value of the domain of X using the $index$ dictionary.

The implementation of $BijectFct(M, S, T)$ is realized through $MapVar(M, S, T) \wedge AllDiffExceptVal(I, \bot) \wedge |S| = |T|$, where I is the array representing the FD variables M_x, and $AllDiffExceptVal$ holds when all the FD variables in I are different when their value is not $\bot$ [32].

Given $MapVar(M, S, T)$, the propagation between M, S and T is based on their relationship described in the previous section, and is achieved by maintaining the following invariants :

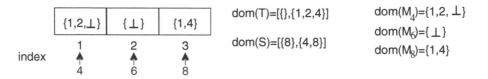

Fig. 1. Implementation of $MapVar(M, S, T)$ (with initial domain $dom(S) = [\{8\}, \{4, 6, 8\}]$ and $dom(T) = [\{\}, \{1, 2, 4\}]$), assuming (other) constraints already achieved some pruning

- $\overline{D}(S) = \{x \mid D(M_x) \neq \{\bot\}\}$
- $\underline{D}(S) = \{x \in \overline{D}(S) \mid \bot \notin D(M_x)\}$
- $\overline{D}(T) = \{v \mid v \neq \bot \wedge \exists x : v \in D(M_x)\}$
- $\underline{D}(T) \supseteq \{v \mid v \neq \bot \wedge \exists x : D(M_x) = \{v\}\}$

The last invariant is not an equality because when a value is known to be in T, it is not always possible to decide which element in I should be assigned to v.

Propagations rules are then easily derived from these invariants (two rules per invariant) :

$$M_x = \bot \rightarrow x \notin \overline{D}(S)$$
$$x \notin \overline{D}(S) \rightarrow M_x = \bot$$
$$x \in \underline{D}(S) \rightarrow M_x \neq \bot$$
$$M_x \neq \bot \rightarrow x \in \underline{D}(S)$$
$$v \notin \overline{D}(T) \wedge v \neq \bot \rightarrow v \notin D(M_x)$$
$$NbOccur(I, v) = 0 \wedge v \neq \bot \rightarrow v \notin \overline{D}(T)$$
$$M_x = v \neq \bot \rightarrow v \in \underline{D}(T)$$
$$v \in \underline{D}(T) \wedge NbOccur(I, v) = 1 \wedge v \in D(M_x) \rightarrow M_x = v$$

where $NbOccur(I, v)$ denotes the number of occurrences of v in the domains of the FD variables in I. Each of these propagation rules can be implemented in $O(1)$ (assuming a bit representation of sets). The implementation of propagators also exploits the cardinality information associated with set variables.

3.5 A Global Constraint Based on Matching Theory

The above propagators do not prune the M_x FD variables (except the $\bot$ value). We show here how flow and matching theory can be used to design a complete filtering algorithm for the $MapVar(M, S, T)$ constraint. The algorithm is similar to that of the GCC and Alldiff constraints but is based on a slightly different notion: the V-matchings (see [33]). In the remainder of this section we show that V-matchings characterize the structure of the $MapVar$ constraint. Note that it also has similarities with the Nvalue, Range and Roots constraints ([34,35]).

Definition 1. *The variable-value graph of a $MapVar(M, S, T)$ constraint is a bipartite graph where the two classes of nodes are the elements of $\overline{D}(S)$ on one side and the elements of $\overline{D}(T)$ plus $\bot$ on the other side. An arc (x, v) is part of the graph iff $v \in D(M_x)$.*

Definition 2. *In a bipartite graph $g = (N_1 \cup N_2, A)$, a matching M is a subset of the arcs such that no two arcs share an endpoint : $\forall (u_1, v_1) \neq (u_2, v_2) \in M : u_1 \neq u_2 \wedge v_1 \neq v_2$. A matching M covers a set of nodes V, or M is a V-matching of g iff $\forall x \in V : \exists (u, v) \in M : u = x \vee v = x$.*

The following property states the relationship between matching in the bipartite graphs and solutions of the $MapVar$ constraint.

Property 1. Given the constraint $MapVar(M, S, T)$ and its associated variable-value graph g, assuming the constraint is consistent, we have :

- (1) Any solution $m : s \rightarrow t$ contains a t-matching of g, and any t-matching can be extended to a solution.
- (2) An arc (x, v) belongs to a $\underline{D}(T)$-matching of g, *iff* there exists a solution m with $m(x) = v$

Proof. (1) The solution m is surjective; every node of t must have at least one incident arc. If we choose one incident arc per node in t, we have a t-matching as m is a function.

Given a t matching, let $m : s \rightarrow t$ be the bijective function corresponding to this matching. Adding arcs to t leads to a surjective function. Let $s' = \underline{D}(S) \cup s$, and $t' = \underline{D}(T) \cup t$. Since the constraint is consistent, $\forall x \in s' \setminus s \; \exists (x, v) \in g : v \neq \perp$, and $\forall v \in t' \setminus t \; \exists (x, v) \in g$. Adding all these arcs leads to a surjective function which is a solution.

(2) ($\Rightarrow$) This is a special case of the second part of (1).

($\Leftarrow$)Let $m : s \rightarrow t$ be a solution with $m(x) = v$. We then have $(x, v) \in g$. By (1), the graph g contains a t-matching M which is also a $\underline{D}(T)$-matching as $\underline{D}(T) \subseteq t$. If $(x, v) \in M$ we are done. Assume $(x, v) \notin M$. Then x is free with respect to M because $M(x) = v$. As $v \in t$, v is covered by M; there is a variable node w such that $(w, v) \in M$. Then, x, v, w is an even alternating path starting in a free node. Replacing (w, v) by (x, v) leads to another t-matching, hence a $\underline{D}(T)$-matching of g.   ∎

From Property 1, an arc-consistency filtering algorithm can be derived : compute the set A of arcs belonging to some $\underline{D}(T)$-matching of the bipartite graph; if $(x, v) \notin A$, remove v from $D(M_x)$. The computation of this set can be done using techniques such as described in [33], with a complexity of $O(mn)$, where n is the size of T, and m is the number of arcs in the variable-value graph.

4   Approximate Graph Matching and Other Matching Problems

In this section, we define different matching problems ranging from graph homomorphism to approximate subgraph matching. The following definitions apply for directed as well as undirected graphs.

A **graph homomorphism** between a pattern graph $P = (N_p, A_p)$ and a target graph $G = (N, A)$ is a total function $f : N_p \rightarrow N$ respecting the *morphism constraint* $(u, v) \in A_p \Rightarrow (f(u), f(v)) \in A$. The graph P is homomorphic to G through the function f. In a **graph monomorphism**, the function f must be injective. In a **graph isomorphism** the function f must be bijective, and the condition $(u, v) \in A_p \Rightarrow (f(u), f(v)) \in A$ is replaced by $(u, v) \in A_p \Leftrightarrow (f(u), f(v)) \in A$. **Subgraph** isomorphisms is defined over an induced subgraph of the target graph. Notice that subgraph homo/mono-morphism are meaningless as graph homo/mono-morphism already maps N_p to a subset of N. All these problems, except graph isomorphism are NP-complete.

A useful extension is *approximate* subgraph matching, where the pattern graph and the found subgraph in the target graph may differ with respect to their structure [9]. We choose an approach where the approximations are declared by the user in the pattern graph through optional nodes and forbidden arcs.

In the previous graph matching problems, all the nodes of the pattern must be matched. An interesting extension consists in allowing optional nodes in the pattern graph. Those nodes need not necessarily be matched. If they are, all arcs incident to them are considered part of the pattern and the matching constraints apply to them. In other words, the pattern that is used in the morphism problem is an induced subgraph of the pattern containing optional nodes.

In graph isomorphism, if two nodes in the pattern are not related by an arc, this absence of arc is an implicit forbidden arc in the matching. It would be interesting to declare explicitly which arcs are *forbidden*, leading to problems between monomorphism and isomorphism.

In Figure 2, mandatory nodes are represented as filled nodes, and optional nodes are represented as empty nodes. Mandatory arcs are represented with plain line, and arcs incident to optional nodes are represented with dashed lines. Forbidden arcs are represented with a plain line crossed.

In that figure, node 6 cannot be matched to node f because only one of the arcs $(6,4)$ and $(6,5)$ in the pattern can be matched in the target. The right side of the figure presents two solutions of the matching problem. The nodes and arcs not matched in the target graph are greyed.

A pattern graph with optional nodes and forbidden arcs forms an *approximate pattern graph*, and the corresponding matching is called an *approximate subgraph matching*[9]. We focus here on approximate graph monomorphism.

Definition 1. *An **approximate pattern graph** is a tuple* (N_p, O_p, A_p, F_p) *where* (N_p, A_p) *is a graph,* $O_p \subseteq N_p$ *is the set of optional nodes and* $F_p \subseteq N_p \times N_p$ *is the set of forbidden arcs, with* $A_p \cap F_p = \emptyset$.

Definition 2. *An **approximate subgraph matching** between an approximate pattern graph* $P = (N_p, O_p, A_p, F_p)$ *and a target graph* $G = (N, A)$ *is a partial function* $f : N_p \to N$ *such that:*

1. $N_p \setminus O_p \subseteq dom(f)$
2. $\forall\, i,j \in dom(f) : i \neq j \Rightarrow f(i) \neq f(j)$
3. $\forall\, i,j \in dom(f) : (i,j) \in A_p \Rightarrow (f(i), f(j)) \in A$
4. $\forall\, i,j \in dom(f) : (i,j) \in F_p \Rightarrow (f(i), f(j)) \notin A$

The notation $dom(f)$ represents the domain of f. Elements of $dom(f)$ are called the selected nodes of the matching. According to this definition, if $F_p = \emptyset$ the matching is a subgraph monomorphism, and if $F_p = N_p \times N_p \setminus A_p$, the matching is an isomorphism.

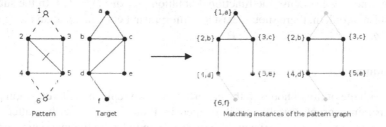

Pattern   Target   Matching instances of the pattern graph

Fig. 2. Example of approximate matching

5   Modeling Graph Matching and Related Problems

In this section, we show how CP(Graph+Map) can be used for modeling and solving a
wide range of graph matching problems.

The problems of graph matching can be stated along three different dimensions:

- homomorphism versus monomorphism versus isomorphism;
- graph versus subgraph matching;
- exact versus approximate matching

These different problems illustrated in Table 1. All these problems can be modeled and
solved through a morphism constraint on a map variable and two graph variables.

5.1   The Basic Morphism Constraints

The two important morphism constraints introduced in this paper are the
$SurjMC(P,G,M)$ and $BijMC(P,G,M)$ constraints, which holds when M is a total surjective / bijective mapping from P to G respecting the morphism constraint.

$$SurjMC(P,G,M) \equiv SurjectFct(M, Nodes(P), Nodes(G)) \wedge MC(P,G,M)$$
$$BijMC(P,G,M) \equiv BijectFct(M, Nodes(P), Nodes(G)) \wedge MC(P,G,M)$$
$$\text{with}\quad MC(P,G,M) \equiv \forall(i,j) \in Arcs(P) : (M(i),M(j)) \in Arcs(G)$$

We now show how these two morphism constraints can be used to solve the different
classes of problems.

5.2   Exact Matching

Let p be a pattern graph and g be a target graph. The graphs p and g are ground objects
in CP(Graph+Map). Graph homo and monomorphism can easily be modeled as shown
in Table 1. Homomorphim (resp. monomorphism) requires a surjective (resp. bijective)
function between p and a subgraph of g, respecting the morphism constraint. We use here
a graph variable instead of a graph constant for the target graph (G with $D(G) = [\emptyset, g]$).

Graph isomorphism requires a bijective function between p and g respecting two
morphism constraints : one between the graphs, and a second between the complementary graphs. This requires a complementary graph constraint $CompGraph(G,Gc)$
which holds if $Nodes(G) = Nodes(Gc) = N$ and $Arcs(Gc) = (N \times N) \setminus Arcs(G)$.
For conciseness, we also use the functional notation $Comp(G) = Gc$. In the subgraph
isomorphism problem, there should exist a isomorphism between p and an induced sub-
graph of g.

5.3   Optional Nodes and Forbidden Arcs

To cope with the optional nodes in the pattern graph, we replace the fixed graph pattern
by a constrained graph variable, as illustrated in Table 1. Let p be the pattern graph
with optional nodes, and p_{man} be the subgraph of p induced by the mandatory nodes of
p. Graph monomorphisms with optional nodes amounts to find an intermediate graph

Table 1. Constraints for the matching problems

Exact matching	
homomorphism	$G \subseteq g \wedge SurjMC(p, G, M)$
monomorphism	$G \subseteq g \wedge BijMC(p, G, M)$
isomorphism	$BijMC(p, g, M) \wedge BijMC(Comp(p), Comp(g), M)$
subgraph isomorph.	$G \subseteq^* g \wedge BijMC(p, G, M) \wedge BijMC(Comp(p), Comp(G), M)$
Optional nodes	
homomorphism	$P \in [p_{man}, p] \wedge P \subseteq^* p \wedge G \subseteq g \wedge SurjMC(P, g, M)$
monomorphism	$P \in [p_{man}, p] \wedge P \subseteq^* p \wedge G \subseteq g \wedge BijMC(P, g, M)$
isomorphism	$P \in [p_{man}, p] \wedge P \subseteq^* p \wedge BijMC(P, g, M)$
	$\wedge BijMC(Comp(P), Comp(G), M)$
subgraph isomorph.	$G \subseteq^* g \wedge P \in [p_{man}, p] \wedge P \subseteq^* p \wedge BijMC(P, G, M)$
	$\wedge BijMC(Comp(P), Comp(g), M)$
Forbidden arcs	
monomorphism	$G \subseteq^* g \wedge BijMC(p, G, M) \wedge BijMC(p_{forb}, Comp(G), M)$

between p_{man} and p which is monomorphic to the target graph. However, between p_{man} and p, only the subgraphs induced by p should be considered. When two optional nodes are selected in the matching, if there is an arc between these nodes in pattern graph p, this arc must be considered in the matching, according to our definition of optional nodes, this is done through the use of the induced subgraph relation ($\subseteq^*$).

When all the nodes of the pattern graph are optional in the graph monomorphism, we have the *maximum common subgraph* problem by adding the size of P as an objective function. Similarly for subgraph isomorphism, this leads to the *maximum common induced subgraph* problem.

Allowing the specification of a set of forbidden arcs amounts to a simple generalization of the isomorphism problem, lying between monomorphism and isomorphism. As in the model for isomorphism, forbidden arcs are handled through a morphism constraint on the complement of the target graph. This time, only a specified set p_{forb} of arcs are forbidden. Isomorphism constitutes a special case where $p_{forb} = Arcs(Comp(p))$. This illustrated for the monomorphism problem in Table 1

The problem of approximate subgraph matching as defined in section 5, simply combines the use of optional nodes and forbidden arcs. Given an approximate pattern graph (N_p, O_p, A_p, F_p) where (N_p, A_p) is a graph, $O_p \subseteq N_p$ is the set of optional nodes, and $F_p \subseteq N_p \times N_p$ is the set of forbidden arcs, and a target graph (N, A), we define the following CP(Graph+Map) constants :

- p: the pattern graph (N_p, A_p),
- p_{man}: the subgraph of p induced by the mandatory nodes $N_p \setminus O_p$ of p,
- g: the target graph (N, A),
- p_{forb} : the graph (N_p, F_p) of the forbidden arcs.

The modeling of approximate matching is then a combination of graph monomorphism with optional nodes, and forbidden arcs.

$$G \subseteq^* g \wedge P \in [p_{man}, p] \wedge P \subseteq^* p \wedge BijMC(P, G, M)$$
$$\wedge Nodes(Pc) = Nodes(P) \wedge Pc \subseteq^* p_{forb} \wedge BijMC(Pc, Comp(G), M)$$

5.4   Global Constraints

The main difference between the $SurjMC(P, G, M)$ and $BijMC(P, G, M)$ constraints is an alldiff constraint ensuring the bijective property of the mapping M. A direct implementation of these constraints based on their definition would be very inefficient. A global constraint for

$$MC(P, G, M) \equiv \forall (i, j) \in Arcs(P) : (M(i), M(j)) \in Arcs(G)$$

has been designed based on [2,9], and generalized in the context of graph intervals and our extension to function variables. This global constraint is *algorithmically* global as it achieves the same consistency as the original conjunction of constraints, but more efficiently [36].

Redundant constraint, such as proposed in [2,9] have also been developed to enhance the pruning. We also specialized global constraints for the different matching families. For instance, a global constraint for filtering subgraph isomorphism was developed and was used to solve difficult instances in [37]. Regarding the approximate matching with optional nodes, the $Mono$ propagator is specialized and assumes that a $P \subseteq^* p$ constraint is posted too, allowing a more efficient pruning. For the isomorphism and for approximate matching with forbidden arcs, a single propagator combining the two $Mono$ propagator is also used, following the ideas developed in [9].

6   Experimental Results

This section assesses the performance of the proposed CP(Graph+Map) framework for graph matching. We compare our proposed solution with vflib [38,39], the current state of the art algorithm for subgraph isomorphism, improving over Ullman's algorithm [40].

The CP(Graph+Map) framework has been implemented over the Gecode system (http://www.gecode.org), including graph variables and propagators, map variables and propagators, together with matching propagators.

Our benchmark set consists of graphs made of different topological structures as explained in [2]. These graphs were generated using the Stanford GraphBase [41], consisting of 1225 undirected instances, and 405 directed instances. The graphs range from 10 to 125 nodes for undirected graphs, and from 10 to 462 for directed graphs.

The experiments consist in performing subgraph monomorphism over the 1225 undirected instances, and subgraph isomorphism over the 405 instances. All solutions are searched. Following the methodology used in [2], we ran the two competing algorithms for five minutes on each of the problem instances. A run is called *solved* if it finishes under five minutes or *unsolved* otherwise. All benchmarks were performed on an Intel Xeon 3 Ghz.

Table 6 shows the experimental results. We report the percentage of solved instances (sol.), the percentage of unsolved instances (unsol), the total running time (tot.T), the mean running time (av.T) and memory (av.M) and the mean running time and memory over instances solved by both approaches (resp. "av.T com." and "av.M com.").

The CP(Graph+Map) model solves more problem instances than the specialized vflib algorithm. This difference is significant for subgraph monomorphism (61% vs.

Table 2. Comparison of the two methods on monomorphism and isomorphism problems

All solutions; subgraph monomorphism over undirected graphs (5 min. limit)							
	solved	unsolved	tot.T min	av.T sec	av.M kb	av.T com. sec	av.M com. kb
vflib	48%	51%	3273	160	11.91	4.96	97.6
CP(Graph+Map)	61%	38%	2479	121	9115.46	5.43	8243
All solutions; subgraph isomorphism over directed graphs (5 min. limit)							
	solved	unsolved	tot.T min	av.T sec	av.M kb	av.T com. sec	av.M com. kb
vflib	92%	7%	181	26.95	114.28	4.11	4.22
CP(Graph+Map)	96%	3%	109	16.22	2859.85	5.04	2754

48%). It is interesting to notice that around 4% of the instances solved by vflib were not solved by our CP model. This shows that on some instances, standard algorithms can be better, but that globally, CP(Graph+Map) solves more instances. It is clear that the CP approach consumes more memory. The comparison of the average time is clearly in favour of CP(Graph+Map) as it solves more instances. It is more interesting to compare the mean execution time on the commonly solved instances. This shows that the time overhead induced by the CP approach is minimal on the commonly solved instances : about 9% for monomorphism over undirected graphs and 22% for isomorphism over directed graphs.

We conclude that our approach is beneficial to someone willing to pay an average time overhead of 9% to 22% on "simple" instances to be able to solve a fourth of the instances of the benchmark which cannot be solved in the time limit by the other method.

7   Conclusion

In this paper, we showed how the integration of two domains of computation over count-able structures, *graphs*[13]. and *maps*, [16], can be used for modeling and solving a wide spectrum of of graph matching problems with any combination of the following properties : monomorphism or isomorphism, graph or subgraph matching, exact or approximate matching (user-specified approximation [9]). To achieve this, we needed to generalize the map variables with non-fixed source and target sets (of the Cardinal kind [31]).

We showed how a single constraint able to use both fixed and non-fixed graph variables is sufficient to model all these graphs matching problems. Furthermore we showed that this constraint programming approach is competitive with the state of the art algorithm for subgraph isomorphism vflib based on the Ullman graph matching algorithm; by solving substantially more instances (our approach solves more complex instances) and requiring a small overhead over the simple instances.

Future work includes the definition of consistency for map variables, the analysis of the impact of our flow-based filtering algorithm for map variables, the design of a more efficient algorithm (we target $O(\sqrt{m}n)$) for this global constraint and the extension of graph matching to other graph comparison problems such as subgraph bisimulation [42].

Acknowledgments

The authors want to thank the anonymous reviewers for the helpful comments. Thanks to Pierre Dupont for his comments on an earlier version of this paper. This research is supported by the Walloon Region, project Transmaze (WIST516207) and by the Interuniversity Attraction Poles Programme (Belgian State, Belgian Science Policy).

References

1. Conte, D., Foggia, P., Sansone, C., Vento, M.: Thirty years of graph matching in pattern recognition. IJPRAI 18, 265–298 (2004)
2. Larrosa, J., Valiente, G.: Constraint satisfaction algorithms for graph pattern matching. Mathematical. Structures in Comp. Sci. 12, 403–422 (2002)
3. Rudolf, M.: Utilizing constraint satisfaction techniques for efficient graph pattern matching. In: Ehrig, H., Engels, G., Kreowski, H.-J., Rozenberg, G. (eds.) TAGT 1998. LNCS, vol. 1764, pp. 238–251. Springer, Heidelberg (1998)
4. Wang, J.T.L., Zhang, K., Chirn, G.W.: Algorithms for approximate graph matching. Inf. Sci. Inf. Comput. Sci. 82, 45–74 (1995)
5. Messmer, B.T., Bunke, H.: A new algorithm for error-tolerant subgraph isomorphism detection. IEEE Trans. Pattern Anal. Mach. Intell. 20, 493–504 (1998)
6. DePiero, F., Krout, D.: An algorithm using length-r paths to approximate subgraph isomorphism. Pattern Recogn. Lett. 24, 33–46 (2003)
7. Robles-Kelly, A., Hancock, E.: Graph edit distance from spectral seriation. IEEE Transactions on Pattern Analysis and Machine Intelligence 27-3, 365–378 (2005)
8. Giugno, R., Shasha, D.: Graphgrep: A fast and universal method for querying graphs. ICPR 2, 112–115 (2002)
9. Zampelli, S., Deville, Y., Dupont, P.: Approximate constrained subgraph matching. In: Verlag, S. (ed.) International Conference on Principles and Practice of Constraint Programming, pp. 832–836 (2005)
10. Sorlin, S., Solnon, C.: A global constraint for graph isomorphism problems. In: Régin, J.-C., Rueher, M. (eds.) CPAIOR 2004. LNCS, vol. 3011, pp. 287–302. Springer, Heidelberg (2004)
11. Puget, J.F.: Symmetry breaking revisited. Constraints 10, 23–46 (2005)
12. Mamoulis, N., Stergiou, K.: Constraint satisfaction in semi-structured data graphs. In: Wallace, M. (ed.) CP 2004. LNCS, vol. 3258, pp. 393–407. Springer, Heidelberg (2004)
13. Dooms, G., Deville, Y., Dupont, P.: Cp(graph): Introducing a graph computation domain in constraint programming. In: Verlag, S. (ed.) International Conference on Principles and Practice of Constraint Programming, pp. 211–225 (2005)
14. Gervet, C.: New structures of symbolic constraint objects: sets and graphs. In: Third Workshop on Constraint Logic Programming (WCLP 1993), Marseille (1993)
15. Chabrier, A., Danna, E., Pape, C.L., Perron, L.: Solving a network design problem. Annals of Operations Research 130, 217–239 (2004)
16. Gervet, C.: Interval propagation to reason about sets: Definition and implementation of a practical language. Constraints 1, 191–244 (1997)
17. Flener, P., Hnich, B., Kiziltan, Z.: Compiling high-level type constructors in constraint programming. In: Ramakrishnan, I.V. (ed.) PADL 2001. LNCS, vol. 1990, pp. 229–244. Springer, Heidelberg (2001)

18. Beldiceanu, N., Contjean, E.: Introducing global constraints in CHIP. Mathematical and Computer Modelling 12, 97–123 (1994)
19. Pesant, G., Gendreau, M., Potvin, J., Rousseau, J.: An exact constraint logic programming algorithm for the travelling salesman with time windows. Transportation Science 32, 12–29 (1996)
20. Puget, J.F.: Pecos a high level constraint programming language. In: Proceedings of Spicis 1992 (1992)
21. Beldiceanu, N., Flener, P., Lorca, X.: The tree constraint. In: CPAIOR, pp. 64–78 (2005)
22. Sellmann, M.: Cost-based filtering for shorter path constraints. In: Rossi, F. (ed.) CP 2003. LNCS, vol. 2833, pp. 694–708. Springer, Heidelberg (2003)
23. Cambazard, H., Bourreau, E.: Conception d'une contrainte globale de chemin. In: 10e Journées nationales sur la résolution pratique de problFmes NP-complets (JNPC 2004), pp. 107–121 (2004)
24. Dooms, G., Katriel, I.: The minimum spanning tree constraint. In: Proceedings of the 12th International Conference on Principles and Practice of Constraint Programming, pp. 152–166 (2006)
25. Dooms, G., Katriel, I.: The "not-too-heavy" spanning tree constraint. In: Van Hentenryck, P., Wolsey, L.A. (eds.) CPAIOR 2007. LNCS, vol. 4510. Springer, Heidelberg (2007)
26. Hnich, B.: Function variables for Constraint Programming. PhD thesis, Uppsala University, Department of Information Science (2003)
27. Frisch, A.M., Jefferson, C., Hernandez, B.M., Miguel, I.: The rules of constraint modelling. In: Proceedings of IJCAI 2005 (2005)
28. Lauriere, J.L.: A language and a program for stating and solving combinatorial problems. Artificial Intelligence 10, 29–128 (1978)
29. Smith, D.: Structure and design of global search algorithms. Technical Report Tech. Report KES.U.87.12, Kestrel Institute, Palo Alto, Calif (1987)
30. Cadoli, M., Palopoli, L., Schaerf, A., Vasile, D.: NP-SPEC: An executable specification language for solving all problems in NP. In: Gupta, G. (ed.) PADL 1999. LNCS, vol. 1551, pp. 16–30. Springer, Heidelberg (1999)
31. Azevedo, F.: Cardinal: A finite sets constraint solver. Constraints 12, 93–129 (2007)
32. Beldiceanu, N.: Global constraints as graph properties on structured network of elementary constraints of the same type. Technical Report T2000/01, SICS (2000)
33. Thiel, S.: Efficient Algorithms for Constraint Propagation and for Processing Tree Descriptions. PhD thesis, University of Saarbrucken (2004)
34. Bessière, C., Hebrard, E., Hnich, B., Kiziltan, Z., Walsh, T.: Filtering algorithms for the nvalue constraint. In: Barták, R., Milano, M. (eds.) CPAIOR 2005. LNCS, vol. 3524, pp. 79–93. Springer, Heidelberg (2005)
35. Bessière, C., Hebrard, E., Hnich, B., Kiziltan, Z., Walsh, T.: The range and roots constraints: Specifying counting and occurrence problems. In: IJCAI, pp. 60–65 (2005)
36. Bessière, C., Van Hentenryck, P.: To be or not to be ... a global constraint. In: Rossi, F. (ed.) CP 2003. LNCS, vol. 2833, pp. 789–794. Springer, Heidelberg (2003)
37. Zampelli, S., Deville, Y., Solnon, C., Sorlin, S., Dupont, P.: Filtering for subgraph isomorphism. In: Proc. 13th Conf. of Principles and Practice of Constraint Programming. LNCS, pp. 728–742. Springer, Heidelberg (2007)
38. Foggia, P., Sansone, C., Vento, M.: An improved algorithm for matching large graphs. In: ed. 3rd IAPR-TC15 Workshop on Graph-based Representations. (2001), http://amalfi.dis.unina.it/graph/db/vflib2.0/doc/vflib.html

39. Cordella, L.P., Foggia, P., Sansone, C., Vento, M.: Performance evaluation of the vf graph matching algorithm. In: ICIAP, pp. 1172–1177. IEEE Computer Society, Los Alamitos (1999)
40. Ullmann, J.R.: An algorithm for subgraph isomorphism. J. ACM 23, 31–42 (1976)
41. Knuth, D.E.: The Stanford GraphBase. A Platform for Combinatorial Computing. ACM Press, New York (1993)
42. Dovier, A., Piazza, C.: The subgraph bisimulation problem. IEEE Transaction on Knowledge and Data Engineering 15, 1055–1056 (2003)

Quasi-Linear-Time Algorithms by Generalisation of Union-Find in CHR

Thom Frühwirth

Faculty of Computer Science
University of Ulm, Germany
www.informatik.uni-ulm.de/pm/mitarbeiter/fruehwirth/

Abstract. The union-find algorithm can be seen as solving simple equations between variables or constants. With a few lines of code change, we generalise its implementation in CHR from equality to arbitrary binary relations. By choosing the appropriate relations, we can derive fast incremental algorithms for solving certain propositional logic (SAT) problems and polynomial equations in two variables. In general, we prove that when the relations are bijective functions, our generalisation yields a correct algorithm. We also show that bijectivity is a necessary condition for correctness if the relations include the identity function.

The rules of our generic algorithm have additional properties that make them suitable for incorporation into constraint solvers: from classical union-find, they inherit a compact solved form and quasi-linear time and space complexity. By nature of CHR, they are anytime and online algorithms. They solve and simplify the constraints in the problem, and can test them for entailment, even when the constraints arrive incrementally.

1 Introduction

Constraint Handling Rules (CHR) [Frü98, FA03, Frü08] is a logical constraint-based concurrent committed-choice programming language consisting of guarded rules that rewrite conjunctions of atomic formulas. The classical optimal union-find algorithm [TvL84] can be implemented in CHR with best-known quasi-linear time complexity [SF06, SF05]. This result is not accidental, since the paper [SSD05] shows that every (RAM machine) algorithm with at least linear time complexity (an algorithm that at least reads all of its input), can be implemented in CHR with best known time and space complexity. Such a result is not known to hold in other pure declarative programming languages.

The *union-find algorithm* maintains disjoint sets under the operation of union. By definition of set operations, a union operator working on representatives of sets is an equivalence relation, i.e. we can view sets as equivalence classes. Especially iff the elements of the set are variables or constants, union can be seen as equating those elements and giving an efficient way of finding out if two elements are equivalent (i.e., in the same set).

This paper investigates the question if the union-find algorithm written in CHR can be generalised so that other relations than simple equations between

F. Fages, F. Rossi, and S. Soliman (Eds.): CSCLP 2007, LNAI 5129, pp. 91–108, 2008.
© Springer-Verlag Berlin Heidelberg 2008

two variables are possible without compromising correctness and efficiency. Our generalised union-find algorithm then maintains relations between elements under the operation of adding relations.

From classical optimal union-find, our generalised algorithm inherits amortised quasi-linear time and space complexity as well as the possibility to both assert relations (tell) and test for entailed (implied) relations (ask) as will be explained below.

CHR is used here as an effective general-purpose programming language for implementing classical algorithms. Still, by nature of CHR, our implementation is an *anytime (approximation) algorithm and online (incremental) algorithm*. Anytime algorithms can be stopped during their execution and exhibit an intermediate result from which to continue computation. In CHR, we can stop after any rule application and observe the intermediate result in the current state (store). Online algorithm means that the input (in CHR, the constraints of a query) can arrive incrementally, one after the other, without the need to recompute from scratch.

In the context of this paper, the algorithms are used in a classical way, i.e. with certain input to produce a desired output. We do not make use of the concurrent constraint programming features of CHR, where execution is also possible with incomplete and unknown inputs and where constraints - in this context, operations - are delayed until further information becomes available.

We can interpret the relations that we maintain as non-trivial equations. Under this point of view, our generalisation produces a *compact solved normal form* that represents all solutions of the given problem (a set of equations). The solution has at most the size of the original problem. We can also test the equations for entailment. All this is possible even when the constraints arrive incrementally. Hence our algorithm and its operation constraints are well-suited to be used inside constraint solvers.

Overview of the Paper. In the next two sections we quickly introduce CHR and then the union-find algorithm and its implementation in CHR. In Section 4 we generalise the union-find algorithm. The next section proves optimal time and space complexity of our generalisation. We then show correctness in Section 6 when the involved relations are bijective functions. We present two instances in Sections 7 and 8. We end with conclusions. This paper is a revised and extended version of the extended abstract [Frü06].

2   Constraint Handling Rules (CHR)

CHR [Frü98, FA03, Frü08] manipulates conjunctions of constraints (relations, predicates, atoms) that reside in a constraint store. In the following, the meta-variables H, G, B and C denote conjunctions of constraints, denoting head (parts of the head), guard, body of a rule and constraints from a state, respectively. In a CHR program, the set of constraints defined in the heads and the set of the constraints used in the guards are disjoint. The former are called CHR constraints. The latter are called built-in constraints and their meaning is defined by a logical theory named CT. Standard built-in constraints are true, false, as well as syntactic equality = and arithmetic equality =:=.

CHR programs are composed of two main types of rules as given in Fig. 1. A third, hybrid kind called simpagation rules is not essential for this paper. In the figure, we also give the declarative, logical reading (meaning) of the rules, where $\bar{y}$ are the variables that appear only in the body B of a rule. W.l.o.g. we assume for simplicity that variables that appear both in the guard and body also appear in the head of a rule. A simplification rule corresponds to a logical equivalence provided the guard holds, while a propagation rule corresponds to an implication.

Simplification rule: $\quad H \Leftrightarrow G \mid B \qquad \forall \bar{x}\,(G \to (H \leftrightarrow \exists \bar{y}\, B))$

Propagation rule: $\quad H \Rightarrow G \mid B \qquad \forall \bar{x}\,(G \to (H \to \exists \bar{y}\, B))$

Fig. 1. Main Types of CHR Rules and their Logical Reading

The *standard (abstract) operational semantics* of CHR is given by a transition system where states are conjunctions of constraints. In Figure 2 we just give the transition for simplification rules, since we only need this kind of rules in this paper. For the propagation rules the transition is very similar, the only difference is that the head is kept. In the transition system, CHR constraints are treated

if $\quad H \Leftrightarrow G \mid B$ is a copy of a rule $H \Leftrightarrow G \mid B$ with new variables $\bar{X}$
and $\quad CT \models \forall (C \to \exists \bar{X}(H{=}H' \wedge G))$
then $(H' \wedge C) \longmapsto (B \wedge H{=}H' \wedge C)$

Fig. 2. State transition for simplification rules

on a syntactic level, while built-in constraints are treated on a semantic level using logic. A simplification rule replaces instances of the CHR constraints H by B provided the guard test C holds. A propagation rule instead just adds B to H without removing anything.

The constraints of the store comprise the *state* of an execution. Starting from an arbitrary initial store (called *query*), CHR rules are applied exhaustively until a fixpoint is reached. The resulting sequence of state transitions is called a *computation*. A rule is applicable, if its head constraints are matched by constraints in the current store one-by-one and if, under this matching, the guard of the rule is logically implied by the constraints in the store. This applicability condition is formally defined by the formula $CT \models \forall (C \to \exists \bar{X}(H{=}H' \wedge G))$ in the transition system, where CT is the logical theory for the constraints used in guards of the rules. Any of the applicable rules can be applied, and the application cannot be undone, it is committed-choice.

The standard semantics is too abstract to describe the details of CHR implementations. For this purpose, an instance of the standard semantics, called *refined semantics* was formalised in [DSdlBH04]. Our complexity proof relies on this refined semantics. We will shortly describe this semantics. It is, however, beyond the scope of the paper to present the details of this semantics.

Queries are executed from left to right and for each new constraint, rules are applied top-down in the textual reading order of the program. Trivial non-termination of propagation rule applications is avoided by applying them at most once to the same constraints. Built-in constraints in the store are simplified and solved, and that in particular variables that are constrained to take a unique value are equated with that value.

In this refined semantics of actual implementations, a CHR constraint in a query can be understood as a procedure that goes efficiently through the rules of the program in the order they are written, and when it matches a head constraint of a rule, it will look for the other, *partner constraints* of the head in the constraint store and check the guard until an applicable rule is found. We consider such a constraint to be *active*. If the active constraint has not been removed after trying all rules, it will be put into the constraint store. Constraints from the store will be reconsidered (woken) if newly added built-in constraints constrain variables of the constraint, because then rules may become applicable since their guards are now implied. In particular this will be the case if a syntactic or arithmetic equality binds a variable to a constant or another variable.

3   The Union-Find Algorithm

In this section we follow the exposition of [SF06]. The classical union-find (also referred to as disjoint-set-union) algorithm was introduced by Tarjan in the seventies [TvL84]. A classic survey on the topic is [GI91]. The algorithm solves the problem of maintaining a collection of disjoint sets under the operation of union. Each set is represented by a rooted tree, whose nodes are the elements of the set. The root is called the *representative* of the set. The representative may change when the set is updated by a union operation. With the algorithm come three operations on the sets:

- make(X): create a new set with the single element X.
- find(X): return the representative of the set in which X is contained.
- union(X,Y): join the two sets that contain X and Y, respectively (possibly changing the representative).

A new element must be introduced exactly once with make before being subject to union and find operations. To find out if two elements are in the same set already, i.e. to check entailment, one finds their representatives and checks them for equality, i.e. checks find(X)=find(Y).

3.1   Implementing Union-Find in CHR

In the naive union-find algorithm without optimisations, the operations are implemented as follows:

- `make(X)`: generate a new tree with the only node X, i.e. X is the root.
- `find(X)`: follow the path from the node X to the root of the tree. Return the root as representative.
- `union(X,Y)`: find the representatives of X and Y, respectively. To join the two trees, we `link` them by making one root point to the other root.

The following CHR program implements the operations and data structures of the naive union-find algorithm. The CHR constraints `make/1`, `union/2`, `find/2` and auxiliary `link/2` define the corresponding operations (functions are written in relational form), so we call them *operation constraints*. The constraints `root/1` and `->/2` (using infix notation) represent the tree data structure and we call them *data constraints*. We use infix notation for `->/2` to evoke the image of a pointer (directed arc).

```
make    @ make(X) <=> root(X).
union   @ union(X,Y) <=> find(X,A), find(Y,B), link(A,B).

findNode @ X -> Y, find(X,R) <=> X -> Y, find(Y,R).
findRoot @ root(X), find(X,R) <=> root(X), R=X.

linkEq  @ link(X,X) <=> true.
link    @ link(X,Y), root(X), root(Y) <=> X -> Y, root(Y).
```

In the rules `findNode` and `findRoot`, the data constraints X->Y and root(X), respectively, occur in the head and body of their rules. In a naive CHR implementation, when the rule applies, they will be removed and immediately re-added again. For the programs discussed in this paper, this causes only constant time overhead. In general, the simpagation rule notation as well as optimising CHR compilers avoid this overhead by leaving the constraint in the store.

3.2 Optimised Union-Find

The basic algorithm requires $\mathcal{O}(n)$ time per find (and union) operation in the worst case, where n is the number of elements (and thus of make operations). With two independent optimisations that keep the tree shallow and balanced, one can achieve logarithmic worst-case and quasi-constant (i.e. almost constant) amortised running time per operation.

The first optimisation is *path compression* for find. It moves nodes closer to the root after a find. After `find(X)` returned the root of the tree, we make every node on the path from X to the root point directly to the root.

The second optimisation is *union-by-rank*. It keeps the tree shallow and balanced by pointing the root of the smaller tree to the root of the larger tree without changing its rank. *Rank* refers to an upper bound of the tree depth (tree height). If the two trees have the same rank, either direction of pointing is chosen and the rank is incremented by one. With this optimisation, the height of the tree can be logarithmically bound.

The following CHR program implements the resulting optimised classical union-find algorithm with path compression for find and union-by-rank [TvL84].

```
make      @ make(X) <=> root(X,0).
union     @ union(X,Y) <=> find(X,A), find(Y,B), link(A,B).

findNode @ X -> Y, find(X,R) <=> find(Y,R), X -> R.
findRoot @ root(X,N), find(X,R) <=> root(X,N), R=X.

linkEq    @ link(X,X) <=> true.
linkLeft @ link(X,Y), root(X,RX), root(Y,RY) <=> RX>=RY |
               Y -> X, root(X,max(RX,RY+1)).
linkRight@ link(X,Y), root(Y,RY), root(X,RX) <=> RY>=RX |
               X -> Y, root(Y,max(RY,RX+1)).
```

When compared to the naive version ufd_basic, we see that root has been
extended with a second argument that holds the rank of the root node. The
union/2 operation constraint is implemented exactly as for the naive algorithm.
The rule findNode has been extended for path compression. By the help of
the variable R that serves as a place holder for the result of the find operation,
path compression is already achieved during the first pass, i.e. during the find
operation. In the body of the rule, the order of constraints find(Y,R), X->R
optimises execution under the refined semantics of CHR, since under left-to-right
execution, the pointer constraint is only introduced when R has been computed
by find. The link rule has been split into two rules, linkLeft and linkRight,
to reflect the optimisation of union-by-rank.

4   Generalised Union-Find

The idea of generalising union find is to replace equations between variables
by binary relations. Our generalised union-find algorithm then maintains relations between elements under the operation of adding relations. The operation
constraints union, find, link and the data constraint -> get an additional argument to hold the relation. The operation union now asserts a given relation
between its two variables, find finds the relation between a given variable and
the root of the tree in which it occurs. The operation link stores the relation in
the tree data constraint. The arcs in the tree are labeled by relations now.

We assume that in queries, all relations are given. While the program in
principle would also compute with unknown relations, we are not interested in
these computations in the context of this paper. Remember that a query contains
only make, union and find operations. A new element must be introduced first
by a single make before subjected to union and find.

We need some standard *operations on relations* from relational algebra and a
non-standard one, combine. The operations are implemented by constraints as
follows, where *id* is the identity function:

- compose(r_1, r_2, r_3) iff $r_3 := r_1 \circ r_2$
- invert(r_1, r_2) iff $r_2 := r_1^{-1}$

- equal(r_1) iff $r_1 = id$
- combine(r_1, r_2, r_3, r_4) iff $r_4 := r_1^{-1} \circ r_3 \circ r_2$

The commutative diagram below shows the relations between the four relations that are arguments of combine. The question mark after AB reminds us that this relation is the one that combine computes from the other three relations.

```
X -- R1 -- A
|          |
R3         R4?
|          |
Y -- R2 -- B
```

The following code extends the CHR implementation of optimal union-find by additional arguments (the relations) and by additional constraints on them (the operations on relations). These additions are in italics for clarity. Our implementation in the Sicstus 3 Prolog CHR library is available at www.informatik.uni-ulm.de/pm/mitarbeiter/fruehwirth/more/ufe.pl and can be run with CHR online at chr.informatik.uni-ulm.de/~webchr/webchr1.

```
make     @ make(X) <=> root(X,0).
union    @ union(X,XY,Y) <=> find(X,XA,A), find(Y,YB,B),
                            combine(XA,YB,XY,AB), link(A,AB,B).


findNode @ X-XY->Y, find(X,XR,R) <=> find(Y,YR,R),
                           compose(XY,YR,XR), X-XR->R.
findRoot @ root(X,N), find(X,XR,R) <=> root(X,N), equal(XR), X=R.


linkEq   @ link(X,XX,X) <=> equal(XX).
linkLeft @ link(X,XY,Y), root(X,RX), root(Y,RY) <=> RX>=RY |
             invert(XY,YX), Y-YX->X, root(X,max(RX,RY+1)).
linkRight@ link(X,XY,Y), root(Y,RY), root(X,RX) <=> RY>=RX |
             X-XY->Y, root(Y,max(RY,RX+1)).
```

The operation constraint union(X,XY,Y) now means that we enforce relation XY between X and Y. The operation find still returns the root for a given node, but also the relation that holds between the node and the root. In the union, combine computes the relation AB that must hold between the roots that are to be linked from the initial relation XY and the relations XA and YB resulting from the two find operations.

Note that in the linkEq rule, the link operation now tests if the relation XX given between two identical variables X is the identity relation. If this is not the case, the overall computation will stop with an inconsistency error. This happens for example, if we try to union the same two variables twice with different incompatible relations between them.

The choice of the operations on relations added in the program is justified by the intended correctness with regard to the logical reading of the program as

discussed in Section 6. Since all relations in a query are given, and since then the
find operation returns a relation, the operations on relations compose, combine
and invert compute a relation (the last argument) from given relations and
equal checks a given relation.

In the body of the rules the sequence of the constraints takes advantage of
the left-to-right execution order of the refined CHR semantics. The order of
constraints is the same as of the corresponding operations in a procedural programming language.

Normalisation and Entailment Checking

By using the find operation, the results can be further normalised: for each
variable in the problem, we issue a find operation. It will return the relation to
the root variable and as a side-effect the tree will be compressed to have a direct
pointer between the two variables. So afterwards, the solved form contains data
constraints of the form X_i-XR->R_j, where all X_i are different (and the R_j are root
variables only).

Also we can *check for entailment*, i.e. ask if a given relation holds between
two given variables. This is the case if their roots are the same (otherwise they
are unrelated) and if asserting the relation using union would not change the
tree. That is, a given relation already holds between two variables if the union
operation leads to a link operation that does not update the tree. This is the
case if the linkEq rule is applicable. Therefore the following special instance of
the union rule suffices for entailment of a relation (X XY Y):

```
unioned?(X,XY,Y) <=>
 find(X,XA,A),find(Y,YB,B), combine(XA,YB,XY,AB), A==B, equal(AB).
```

A==B checks if A and B are identical and equal(AB) checks if AB is the identity
relation. These checks are inherited from the linkEq rule.

The query associated with the entailment test can also be modified to find
out what relation between two given variables X and Y holds. In that case we
replace combine(XA,YB,XY,AB) according to its definition so that it computes
XY from XA, YB and AB (which must be identity). We get the rule:

```
related?(X,XY,Y) <=>
 find(X,XA,A),find(Y,YB,B), A==B, invert(YB,BY),compose(XA,BY,XY).
```

where just X and Y are given.

5 Complexity

Our algorithm is a canonical extension (proper generalisation) of the optimised
union-find algorithm in CHR: we added arguments holding the relations to existing CHR constraints. In the rules, these additional arguments for the relations
are variables. In the head of each rule, these variables are all distinct. The guards
have not been changed. The additional constraints in the rule bodies only involve

variables for relations. These operations on relations can be performed in constant time. Moreover, if we specialise our algorithm to the case where the only relation is identity id, we get back the original program. These observations are an indication that we can preserve the complexity and correctness results of the original algorithm under certain conditions.

We first have to establish some general lemmas for CHR rule applications. We restrict ourselves to simplification rules in this section, since this is the only type of rules we use in this paper. Then, our proof is based on a mapping from computations in our generalised algorithm to computation in the original union-find algorithm.

Definition 1. Given a CHR simplification rule $R = H \Leftrightarrow G \mid B$, then $(H \wedge G)$ is called the *minimal state of R* and the transition $(H \wedge G) \longmapsto (B \wedge G)$ is called the *minimal transition of R*.

Minimal transitions capture the essence of a CHR rule application.

Lemma 1 (Minimal Transitions). A CHR simplification rule $R = H \Leftrightarrow G \mid B$ is applicable to its minimal state, $(H \wedge G)$, leading to the minimal transition of R, $(H \wedge G) \longmapsto (B \wedge G)$.

Proof. The proof is straightforward from the operational semantics of CHR given as transition system in Figure 2. We take a copy of the rule R, $H' \Leftrightarrow G' \mid B'$. It obviously satisfies the rule applicability condition of the transition, $CT \models \forall(G \rightarrow \exists \bar{X}(H'=H \wedge G'))$, where $H'=H$ is simply a variable renaming that makes G and G' equivalent. We can apply this variable renaming in the resulting state $(B' \wedge H'=H \wedge G)$ to get $(B \wedge G)$. □

Lemma 2 (Arbitrary Transitions). Any transition $(H' \wedge C) \longmapsto (B \wedge H=H' \wedge C)$ resulting from application of a rule R (cf. Fig. 2) can be derived from the minimal transition $(H \wedge G) \longmapsto (B \wedge G)$ of R (cf. Lemma 1) by instantiating variables according to the equation $H=H'$ and by replacing the guard constraints G with $H=H' \wedge C$.

Proof. Based on the transition system in Fig. 2, we only have to note that the source states $(H \wedge H=H' \wedge C)$ and $(H' \wedge C)$ are identical: By definition H and $(H' \wedge C)$ do not have any variables in common, therefore we can replace H by H' and remove the redundant resulting equation $H'=H'$. □

In our algorithm, rule applications take constant time as in the original union-find algorithm in CHR if the additional operations on relations take constant time.

Lemma 3 (Union-Find Rule Application Complexity). Every rule of the generalised union-find algorithm can be applied in constant time and space under the refined semantics of CHR [DSdlBH04], if the introduced operations on relations (combine, compose, equal, invert) take constant time and space.

Proof Sketch. The proof is essentially the same as that for the original optimal union-find algorithm as presented in Section 6 of [SF06]. We only repeat the main points here.

Following the refined semantics of CHR [DSdlBH04], CHR implementations exist where all of the following take constant time [SF06]:

- finding all constraints with a particular value in a given argument position (due to indexing),
- matching of constants and variables in the head of a rule,
- testing and solving simple built-in constraints (like =, =< and >=),
- adding and deleting CHR constraints.

It is further assumed that storing the (data) constraints and their indexes takes constant space per constraint and variable.

From the above assumptions it is shown that processing a data constraint under the refined semantics takes constant time: the constraint is called, some rules are tried, some partner constraints which share a variable with the active constraint are looked for, but none are present, and finally the call ends with inserting the data constraint into the constraint store. It follows that all rule tries and applications with an active constraint take constant time.   □

We now can show optimal time and space complexity of our algorithm.

Theorem 1 (Optimal Complexity). Our generalised union-find algorithm in CHR has the same time and space complexity as the original optimised union-find algorithm if the introduced operations on relations (`combine`, `compose`, `equal`, `invert`) take constant time and space.

Proof Sketch. We prove the Theorem by showing that any computation in our generalised algorithm can be mapped into a computation of the original union-find algorithm with the same time and space complexity. The claim is shown by induction on length of the computation and case analysis of the rules applicable in a computation step. It is thus sufficient to consider individual rule applications. Since each rule application takes constant time and space in both algorithms by Lemma 3, it suffices to show that the computations lengths are linearily related.

We construct a function that maps transitions of our generalised union-find algorithm to transitions of the optimised union-find algorithm. The mapping function τ simply removes the additional arguments holding the relations and additional constraints for the operations on the relations. The function τ is exhaustively defined by the following equalities:

$\tau(\,(\texttt{A} \longmapsto \texttt{B})\,) = \tau(\texttt{A}) \longmapsto \tau(\texttt{B})$
$\tau(\,(\texttt{A} \wedge \texttt{B})\,) = \tau(\texttt{A}) \wedge \tau(\texttt{B})$
$\tau(\texttt{make(X)}) = \texttt{make(X)}$
$\tau(\texttt{union(X,XY,Y)}) = \texttt{union(X,Y)}$
$\tau(\texttt{find(X,XY,Y)}) = \texttt{find(X,Y)}$
$\tau(\texttt{link(X,XY,Y)}) = \texttt{link(X,Y)}$
$\tau(\texttt{X=Y}) = \texttt{true}$ if X and Y are relations
$\tau(\texttt{X=Y}) = \texttt{X=Y}$ if X and Y are elements
$\tau(\texttt{X>=Y}) = \texttt{X>=Y}$
$\tau(\texttt{root(X,Y)}) = \texttt{root(X,Y)}$

τ(X-XY->Y)) = X->Y
τ(combine(XA,YB,XY,AB)) = true
τ(compose(XY,YR,XR)) = true
τ(equal(XX)) = true
τ(invert(XY,YX)) = true

Note that the constraint true is the neutral element for conjunction, i.e. true $\wedge$ A and A $\wedge$ true are each the same as A.

We now establish correctness of the mapping: in our generic algorithm, transitions can be caused by the application of the union-find rules or of rules that define the operations on relations (combine, compose, equal, invert) for the specific instance. For the union-find rules, we have to show that for each rule application in a transition of our generic algorithm, there is an application of the rule by the same name in the corresponding mapped transition in the original union-find program.

We need not apply rules to arbitrary states, but just have to consider minimal transitions by Lemma 2. It remains to show that the function τ correctly maps the queries and the minimal transitions for each rule of union-find. For the union rule we have that

τ(union(X,XY,Y) $\longmapsto$ find(X,XA,A) $\wedge$ find(Y,YB,B) $\wedge$
                          combine(XA,YB,XY,AB) $\wedge$ link(A,AB,B)) =
union(X,Y) $\longmapsto$ find(X,A) $\wedge$ find(Y,B) $\wedge$ link(A,B).

Analogously, the mapping can be applied to the other rules of the program.

If one of the rules for operations on relations is applied in a transition, the source and target state in the mapped transition are identical, because τ maps these operations and the built-in syntactic equalities for relations resulting from them all to true. Since we required that these operations take constant time and space, each operation can only cause a constant number of transitions. Therefore their execution causes only a constant time overhead for each rule of our algorithm.

We also have to consider inconsistency errors caused by execution of equal caused by the application of a linkeq rule. In that case, the computation stops in our generic algorithm, while in the corresponding mapped transition, equal is mapped to true and the computation can possibly proceed. Clearly it needs less time and space to stop a computation early.   □

6   Correctness

By correctness of a program we mean that the logical reading of the rules of a program is a logical consequence of a specification given as a logical theory. Since our generalised union-find algorithm maintains relations between elements under the operation of adding relations, the specification is a theory for these relations. Since our program should work with arbitrary relations, we expect the logical reading of its rules to follow from the empty theory, i.e. to be tautologies. We will see that this is not the case for all rules. To this end, we prove that when the relations involved are bijective functions, our generalisation yields a correct algorithm. We also show that bijectivity is a necessary condition for correctness if the relations include the identity function.

In the logical reading of our rules, we replace `union`, `find`, `link` and `->` as intended by the binary relations between their variables (using infix notation), and the constraints for operations on relations by their definitions using functional notation. As usual, formulas are assumed to be universally closed. Even though the logical reading of union-find does not reflect the intended meaning of the `root` data constraint [SF05], the logical reading suffices for our purposes.

```
(make)      make(X) ⇔ root(X,0).
(union)     (X XY Y) ⇔ ∃XA,A,YB,B,AB ((X XA A) ∧ (Y YB B) ∧
                        XA^-1∘XY∘YB=AB ∧ (A AB B))

(findNode)  (X XY Y) ∧ (X XR R) ⇔ ∃YR ((Y YR R) ∧
                        XY∘YR=XR ∧ (X XR R))
(findRoot)  root(X,N) ∧ (X XR R) ⇔ root(X,N) ∧ XR=id ∧ X=R

(linkEq)    (X XX X) ⇔ XX=id
(linkLeft)  RX>=RY ⇒ ((X XY Y) ∧ root(X,RX) ∧ root(Y,RY) ⇔
                ∃YX (XY^-1=YX ∧ (Y YX X) ∧ root(X,max(RX,RY+1))))
(linkRight) RY>=RX ⇒ ((X XY Y) ∧ root(Y,RY) ∧ root(X,RX) ⇔
                (X XY Y) ∧ root(Y,max(RY,RX+1)))
```

Most rules lead to formulas that do not impose any restriction on the binary relations involved. However, the logical reading of `linkEq` and `findRoot` implies that the only relation that is allowed to hold between identical variables is the identity function id. Most importantly, the meaning of the `findNode` rule is a logical equivalence, that is not a tautology and restricts the involved relations. For example, it does not hold for ≤=XR=YR=XY even though ≤ ∘ ≤=≤.

We now show that our implementation is *correct* if the involved relations are *bijective functions*. In that case, the composition operation is precise enough in that it allows to derive any of the three involved relations from the other two. For most other types of relations, our generalised union-find algorithm is not correct, since it looses information due to composition.

Definition 2. A function f is *bijective* if the function is injective and surjective, i.e. $f(\bar{x}) = y \land f(\bar{u}) = v \land (\bar{x} = \bar{u} \lor y = v) \rightarrow \bar{x} = \bar{u} \land y = v$.

Thus a unary function f is bijective if for every x there is exactly one y and vice versa such that $f(x) = y$. Bijective functions are closed under inverse and composition.

Theorem 2 (Correctness). The logical reading of the rules of our generalised union-find algorithm is a consequence of a theory for the relations if these relations are bijective functions.

Proof. The identity function id needed by rules findRoot and linkEq is a bijective function. The findNode rule leads to the non-tautological formula,

(X XR R) ∧ (X XY Y) ⇔ (X XR R) ∧ (Y YR R) where XY∘YR=XR.

This condition is obviously satisfied if the involved relations are bijective functions, because then, for any value given to one of the variables, the values for the other two variables are uniquely determined on both sides of the logical equivalence and there cannot be another triple of values (x, y, z) that has any of the values in the same component. All other rules are tautologies.   □

Next we show that when the identity function id is one of the relations, then all relations must be bijective.

Theorem 3 (Bijectiveness). The logical reading of the rules of our generalised union-find algorithm implies that all relations are bijective if the allowed relations include the identity function.

Proof. In the formula for rule findNode,

(X XR R) ∧ (X XY Y) ⇔ (X XR R) ∧ (Y YR R) where XY∘YR=XR,

we consider two cases in which we replace either relation XY or relation YR by the identity function id. This leads to the two formulas

(X XR R) ∧ (X id Y) ⇔ (X XR R) ∧ (Y XR R) and
(X XR R) ∧ (X XR Y) ⇔ (X XR R) ∧ (Y id R).

The former formula means that any relation XR used must be surjective, the latter means that any relation XR must be injective. Hence any relation must be bijective.   □

The two instances of our generalised union-find algorithm that we will discuss next involve bijective functions only.

7   Instance of Boolean Equations

With our generalised union-find algorithm, we can solve inequations between Boolean variables (propositions), i.e. certain 2-SAT problems. This instance features thus a (small) finite domain and a finite number of relations. In the CHR implementation, the relations are eq for = and ne for ≠, and the truth values are 0 for false and 1 for true. Note that the ne relation holds if the Boolean exclusive-or function (xor) returns true. The operations on relations can be defined by the following rules:

```
compose(eq,R,S) <=> S=R.              invert(R,S) <=> S=R.
compose(R,eq,S) <=> S=R.
compose(R,R,S) <=> S=eq.              equal(S) <=> S=eq.

combine(XA,YB,XY,AB) <=>
    compose(XY,YB,XB), invert(XA,AX), compose(AX,XB,AB).
```

Here is a simple example of a query for Booleans. Note that we introduce the truth values 0 and 1 by `make` and add `union(0,ne,1)` to enforce that they are distinct. This suffices to solve this type of Boolean inequations.

```
?- make(0),make(1),union(0,ne,1),
   make(A),make(B),union(A,eq,B),union(A,ne,0),union(B,eq,1).
root(A,2), B-eq->A, 0-ne->A, 1-eq->A.
```

The result of the query shows that A is also equal to 1. More examples are available online.

Related Work. It is well known that 2-SAT (conjunctions of disjunctions of at most two literals) [APT79] and Horn-SAT (conjunctions of disjunctions with at most one positive literal, i.e. propositional Horn clauses) [BB79, DG84, Min88] can be checked for satisfiability in linear time. The class of Boolean equations and inequations we can deal with is a proper subset of 2-SAT, but not of Horn-SAT, since A ne B $\Leftrightarrow$ (A $\lor$ B) $\land$ ($\neg$A $\lor$ $\neg$B).

These two classical linear-time SAT algorithms are not incremental. They assume that the problem and its graph representation are initially known, because it has to be traversed along its edges. The algorithms only check for satisfiability and can report one possible solution, but they do not simplify or solve the given problem in a general way, so the results are less informative than ours.

The 2-SAT algorithm translates a given problem into a directed graph where arcs are the implications that are logically equivalent to the individual clauses in the problem. It then relies on a linear-time preprocessing of the graph to find is maximal strongly connected components in reverse topological order. Respecting the topological order, truth values are propagated through the components, where all nodes in a component are assigned to the same truth value.

In contrast, our generalised union-find algorithm produces a simple normal form representing all solutions. Due to the properties of the generalised union-find algorithm, our Boolean instance can be integrated into a Boolean constraint solver. For example, the classical Boolean solver in CHR is based on value (unit) propagation, with rules such as `and(X,Y,Z) <=> X=0 | Z=0`, and propagation of equalities, e.g. `and(X,Y,Z) <=> X=Y | Y=Z`. It can be now extended by propagation of inequalities, e.g. `and(X,Y,Z) <=> X ne Y | Z eq 0` and `and(X,Y,Z) <=> X ne Z | X eq 1, Y eq 0, Z eq 0`.

Can we extend our algorithm instance of generalised union-find to deal with 2-SAT? As put to use in the classical algorithm, any disjunction in two variables, A $\lor$ B can be written as implication $\neg$A $\rightarrow$ B. Since we can implement negation using an auxiliary variable, e.g. A ne negA, we just would have to introduce the relation $\rightarrow$ (that corresponds to a total non-strict order $\leq$ on the truth values). But the implication relation looses too much information when composed. For example, given a tree B-$\leq$->A, C-$\leq$->A, B and C can be arbitrarily related. If one now asserts `union(B,eq,C)`, it has no effect on the tree, and thus the information that B eq C is lost.

8   Instance of Linear Polynomials

Another instance of our generalised union-find algorithm deals with linear polynomial equations in two variables. It features an infinite domain and an infinite number of relations. In this instance, the CHR data constraint X-A#B->Y (with A≠0) means X=A*Y+B. The operations on relations are defined as follows:

```
compose(A#B,C#D,S) <=> S = A*C # A*D+B.
invert(A#B,S) <=> S = 1/A # -B/A.          equal(S) <=> S = 1#0.


combine(XA,YB,XY,AB) <=>
            compose(XY,YB,XB), invert(XA,AX), compose(AX,XB,AB).
```

Again, a small example illustrates the behaviour of this instance.

```
?- make(X),make(Y),make(Z),make(W),
   union(X,2#3,Y),union(Y,0.5#2,Z),union(X,1#6,W).
root(X,1), Y-0.5#(-1.5)->X, Z-1.0#(-7.0)->X, W-1.0#(-6.0)->X.
```

Note that the generic linkEq rule asserts that the relation XX in link(X,XX,X) must be the identity function. Thus link(X,1#0,X) is fine, but all other equations of the form link(X,A#B,X) with A#B different from 1#0 will lead to an inconsistency error. While this is correct for link(X,1#1,X), the equation link(X,2#1,X) should not fail as it does, since it has the solution X=-1. Indeed, in our program, an inconsistency will occur whenever a variable is fixed, i.e. determined to take a unique value. Our implementation succeeds exactly when the set of equations has infinitely many solutions.

We now introduce concrete numeric values and solve for determined variables. We express numbers as multiples of the number 1. To make sure that the number 1 always stays the root, so that it can be always found by the find operation, we add root(1,∞) (instead of make(1)) to the beginning of a query.

We split the linkEq rule into two rules. The first restricts applicability of the generic linkEq rule to the case where A=1, the second rule applies otherwise, i.e. to equations that determine their variable (A=\=1) and normalises the equation such that the coefficient is 1 and the second occurrence of the variable is replaced by 1.

```
linkEq1 @ link(X,A#B,X) <=> A=:=1 | B=:=0.
linkEq2 @ link(X,A#B,X) <=> A=\=1 | link(X,1#B/(1-A)-1,1).
```

Note that there is a subtle point about these two rules: X may be the value 1, and in that case the execution of link(X,1#B/(1-A)-1,1) in the right hand side of rule linkEq2 will use rule linkEq1 to check if B/(1-A)-1 is zero (which holds if B = 1-A).

The following small examples illustrate the behaviour of the two new rules (∞ is chosen to be 9):

```
?- root(1,9), make(X),make(Y), union(X,2#3,Y),union(X,4#1,1).
root(1,9), X-4#1->1, Y-0.5#(-1.5)->X.
```

```
?- root(1,9), make(X),make(Y), union(X,4#1,1),union(X,2#3,Y).
root(1,9), X-4#1->1, Y-2#(-1)->1.
```

The queries and answers mean the same, but the answers are syntactically different due to the different order of union operations in the query.

We add another rule that propagates values for determined variables down the tree data structure and so binds all determined variables in linear time:

```
X-A#B->N <=> number(N) | X=A*N+B.
```

```
?- root(1,9), make(X),make(Y), union(X,2#3,Y),union(X,4#1,1).
root(1,9), X=5, Y=1.
```

More examples are available online.

Related Work. [AS80] gives a linear time algorithm that is similar to ours, but is more complicated. Equations correspond to directed arcs in a graph. Like the 2-SAT algorithm [APT79], it computes maximal strongly connected components, and thus the problem has to be known form the beginning. Inside each component, a modification of any linear-time spanning tree algorithm can be used to simplify the equations. The overall effect is the same as with our algorithm, and the algorithm is similar on the components, especially if Kruskal's algorithm [Kru56] for spanning trees is used which relies on union-find. However, our algorithm is simpler and more general in its applicability. It does not need to compute strongly connected components or spanning tress, it directly uses union-find and moreover is incremental.

9   Conclusions

We systematically extended the applicability of union-find algorithm as implemented in CHR. We saw that the generalisation of the algorithm from maintaining equalities to certain binary relations (in particular bijective functions that admit precise composition) is straightforward in CHR and that the generalisation does not compromise quasi-linear time and space efficiency. We have implemented the generalisation and two instances, for equations and inequations over Booleans and for linear polynomial equations in two variables. While linear-time algorithms are known to check satisfiability and to exhibit certain solutions of these problems, our algorithms are simple instances of our generic algorithm. Our implementation in the Sicstus 3 Prolog CHR library is available at `www.informatik.uni-ulm.de/pm/mitarbeiter/fruehwirth/more/ufe.pl` and can be run at `chr.informatik.uni-ulm.de/~webchr/webchr1`.

Our generic algorithm has desirable properties that make it suitable for incorporating its instances into constraint solvers: by nature of CHR, our implementation is an anytime algorithm and online algorithms. The rules solve and simplify the constraints in the problem, and can test them for entailment, even when the constraints arrive incrementally, one after the other.

From classical optimal union-find, our generic algorithm inherits amortised quasi-linear time and space complexity as well as the possibility to both assert relations and test for entailed relations. It produces a compact solved normal form that represents all solutions of the given problem and has at most the size of the original problem. By using the find operation, the results can be further normalised in quasi-linear time. The relation between two given variables can be found in quasi-constant time using find operations.

We have proven that when the relations involved are bijective functions, our generalisation yields a correct algorithm. We also showed that bijectivity is a necessary condition for correctness if the relations include the identity function. While bijective functions may seem quite a strong restriction we remind the reader that permutations, isomorphisms and many other mappings (such as encodings in cryptography) are bijective functions. Indeed, for a domain of size n, there exist $n!$ different bijective functions, i.e. more than exponentially many. Also, most arithmetic functions are at least piecewise bijective, since they are piecewise monotone.

Future work will try to extend the class of bijective functions to other binary relations by abandoning the identity function, and investigate the relationship with classes of tractable constraints. We also would like to find out about the potential tradeoff between efficiency and precision (i.e. when applying our generalised union-find to inequalities like $\leq$).

References

[APT79]   Aspvall, B., Plass, M.F., Tarjan, R.E.: A linear time algorithm for testing the truth of certain quantified Boolean formulas. Information Processing Letters 8, 121–123 (1979)

[AS80]   Aspvall, B., Shiloach, Y.: A fast algorithm for solving systems of linear equations with two variables per equation. Linear Algebra and its Applications 34, 117–124 (1980)

[BB79]   Beeri, C., Bernstein, P.A.: Computational problems related to the design of normal form relational schemas. ACM Trans. Database Syst. 4(1), 30–59 (1979)

[DG84]   Dowling, W.F., Gallier, J.H.: Linear-time algorithms for testing the satisfiability of propositional horn formulae. J. Log. Program. 1(3), 267–284 (1984)

[DSdlBH04]   Gregory, J., Duck, P.J.: Stuckey, Maria Garcia de la Banda, and Christian Holzbaur. In: Demoen, B., Lifschitz, V. (eds.) ICLP 2004. LNCS, vol. 3132. Springer, Heidelberg (2004)

[FA03]   Frühwirth, T., Abdennadher, S.: Essentials of Constraint Programming. Springer, Heidelberg (2003)

[Frü98]   Frühwirth, T.: Theory and Practice of Constraint Handling Rules, Special Issue on Constraint Logic Programming. Journal of Logic Programming 37(1–3), 95–138 (1998)

[Frü06]   Frühwirth, T.: Deriving linear-time algorithms from union-find in chr. In: Schrijvers, T., Frühwirth, T. (eds.) Third Workshop on Constraint Handling Rules, Venice, Italy (July 2006)

[Frü08] Frühwirth, T.: Constraint Handling Rules. Cambridge University Press, Cambridge (to appear, 2008)

[GI91] Galil, Z., Italiano, G.F.: Data Structures and Algorithms for Disjoint Set Union Problems. ACM Comp. Surveys 23(3), 319ff (1991)

[Kru56] Joseph, B., Kruskal, J.B.: On the shortest spanning subtree of a graph and the traveling salesman problem. Proceedings of the American Mathematical Society 7, 48–50 (1956)

[Min88] Minoux, M.: LTUR: a simplified linear-time unit resolution algorithm for Horn formulae and computer implementation. Information Processing Letters 29(1), 1–12 (1988)

[SF05] Schrijvers, T., Frühwirth, T.: Analysing the CHR Implementation of Union-Find. In: 19th Workshop on (Constraint) Logic Programming (W(C)LP 2005), Ulmer Informatik-Berichte 2005-01, University of Ulm, Germany (February 2005)

[SF06] Schrijvers, T., Frühwirth, T.: Optimal union-find in constraint handling rules, programming pearl. Theory and Practice of Logic Programming (TPLP) 6(1) (2006)

[SSD05] Sneyers, J., Schrijvers, T., Demoen, B.: The Computational Power and Complexity of Constraint Handling Rules. In: Second Workshop on Constraint Handling Rules, at ICLP 2005, Sitges, Spain (October 2005)

[TvL84] Tarjan, R.E., van Leeuwen, J.: Worst-case Analysis of Set Union Algorithms. J. ACM 31(2), 245–281 (1984)

Preference-Based Problem Solving for Constraint Programming

Ulrich Junker

ILOG
1681, route des Dolines
06560 Valbonne
France
ujunker@ilog.fr

Abstract. Combinatorial problems such as scheduling, resource allocation, and configuration may involve many attributes that can be subject
of user preferences. Traditional optimization approaches compile those
preferences into a single utility function and use it as the optimization
objective when solving the problem, but neither explain why the resulting solution satisfies the original preferences, nor indicate the trade-offs
made during problem solving. We argue that the whole problem solving
process becomes more transparent and controllable for the user if it is
based on the original preferences. We will use the original preferences
to control this process and to produce explanations of optimality of the
resulting solution. Based on this explanation, the user can refine the
preference model, thus gaining full control over the problem solver.

1   Introduction

Although impressive progress has been made in solving combinatorial optimization problems such as scheduling, resource allocation, and configuration, modern
optimization methods lack easy and wide acceptance in industry. Expert users
often want a detailled control over the choice of a solution and may be reluctant
to give control to an optimizer which ignores existing practices. A good example
is that of planning a project which consists of several tasks. The tasks have to be
assigned to team members and scheduled in time such that constraints on skill,
workload, and task order are respected. There may be standard practices such as
that of assigning team members to their favorite tasks. Those choices may only
be abandoned if they are infeasible or in conflict with more important choices.
If a project planning system completely ignores these standard practices, then
planning experts may be reluctant to accept a plan produced by this system
even if it minimizes global objectives such as the overall project duration.

As standard choices can be abandoned if there is a clear reason, they do
not have the character of constraints, but that of preferences. These preferences
are not formulated on global properties of the resulting plan such as the project
duration, but concern the individual choices that constitute the plan. There may
be preferences for assigning a team member to a task, preferences for planning

F. Fages, F. Rossi, and S. Soliman (Eds.): CSCLP 2007, LNAI 5129, pp. 109–126, 2008.
© Springer-Verlag Berlin Heidelberg 2008

a task in a certain time period and so on. According to the multi-attribute utility theory (MAUT) [9], those individual preferences can be aggregated into a single numerical utility function, which is a weighted sum of subutility functions. This additive utility function fits well into a mixed integer programming (MIP) approach and can be supplied as a maximization objective to a MIP optimizer.

The quality of a MIP solution is then measured with respect to the utility function, but not with respect to the original preferences. For example, a degree of optimality of 99.8% means that as many preferences as possible have been satisfied except for 0.2%. Unfortunately, this explanation does not make sense to the end users who specified the original preferences. If the solution does not give the most preferred choices to the end users, they would like to know whether these most preferred choices are infeasible or whether they are in conflict with other choices that have been made. Explanations of the trade-offs may help users to accept the solution or to revise the preferences.

We therefore argue that solutions of optimization methods need to be enhanced by explanations of optimality in order to become acceptable for the users. Moreover, these explanations should be given in terms of the original preferences and in a form that is comprehensible for non-optimization specialists. Let us consider a simple configuration problem, namely that of choosing a vacation destination. Suppose that Hawaii is preferred to Florida for doing windsurfing, but that Florida has been selected in the solution. The explanation may be that the option Hawaii is infeasible since it is too far away. Or the choice of Hawaii may lead to high hotel costs and thus penalize subsequent choices. Or the choice of Hawaii may be in conflict with more important choices such as visiting a theme park. Explanations of this kind exhibit the trade-offs and the importance orderings that generated the solution and that justify it.

There are other problems of the MAUT approach. Firstly, it assumes that the individual attributes have complete preference orders. If the user has only specified a partial ordering, this order will implicitly be completed when compiling the preferences into a utility function. For example, the user may prefer Hawaii to Florida, Hawaii to the French Riviera, and the French Riviera to Mexico. We may compile this into a utility of 3 for Hawaii, 2 for the French Riviera and for Florida and 1 for Mexico. This implies that Florida is preferred to Mexico, although the user has not stated this. If the options Hawaii and French Riviera are not possible, then only the option Florida is considered and the option Mexico is discarded, although the user might express a preferences for Mexico if both alternatives were presented to her. Hence, implicitly chosen preferences are problematic if they are in conflict with true user preferences. Secondly, if the MAUT approach ranks several solutions in the same way, but the user is not indifferent w.r.t. those solutions, then weight adjustments have to be made in order to differentiate the solutions and to achieve that trade-offs are made in the expected way. As small changes in weights can have a tremendous impact on the result, this process is difficult to achieve manually. Thirdly, there are solutions that represent valid trade-offs, although they do not belong to the

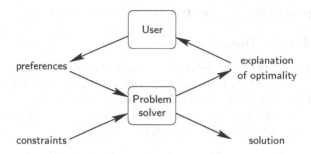

Fig. 1. An interactive optimization process driven by explanations and preferences

convex hull of the solution space. Those solutions cannot be determined by a MAUT approach whatever weights are chosen. As these solutions may represent better compromises than the MAUT-solutions (such as the leximin-optimal solutions in [3]), this limitation is a severe draw-back. Finally, additive utility functions suppose complete preferential independence and are not able to deal with context-dependent preferences [1].

Although the completeness and independence assumptions of the MAUT model are of great benefit for the optimizer, they may hinder fruitful interactions between the user and the optimizer. The user may start with rather incomplete preferences and refine them dependent on the initial solutions. In order to enable this, we follow the vision in [2] and use the original preferences to control the problem solving process. Our approach is based on multi-objective optimization. The problem is decomposed into alternative sequences of single-criterion optimization problems, which can be solved by standard optimizers. The chosen sequence gives information that explains the optimality of the solution. Based on the explanation, the user can either accept the solution or modify the preferences. The problem solver then modifies the solution correspondingly. Preferences thus allow the user to interact with the problem solver and to control its behaviour (see Figure 1). This approach has been applied to configuration problems [8], but is of interest for constraint programming in general.

The paper is organized as follows. We first introduce constraint satisfaction problems with preferences and then discuss different optimization problems. We start with atomic single-criterion optimization problems. Lexicographic optimization tells us how to apply multiple atomic optimization steps for different criteria in sequence. Alternative sequences lead to different lexicographic solutions. The user can choose among them by imposing an importance ordering between criteria. Finally, we discuss Pareto-optimal solutions which represent the different ways to make trade-offs or compromises between conflicting criteria. For each of these optimization problems, we give a solved form and an explanation of optimality which helps the user to modify the preference model and the problem solver outcome. Algorithms for effectively computing optimal solutions are beyond of the scope of this paper and can be found in [5,7].

2   Combinatorial Problems with Preferences

2.1   Variables and Domains

Throughout this paper, we consider a finite set of variables $\mathcal{X}$ where each variable $x \in \mathcal{X}$ has a domain $D(x)$. For example, consider three variables x_1, x_2, x_3 of a vacation configuration example. The domain of x_1 contains the possible activities of the vacation, the domain of x_2 contains the possible vacation destinations, and the domain of x_3 contains the possible hotel chains:

$$D(x_1) := \{\text{Casino, CliffDiving, FilmStudio, SeaPark, WindSurfing}\}$$
$$D(x_2) := \{\text{Acapulco, Antibes, Honolulu, LosAngeles, Miami}\}$$
$$D(x_3) := \{\text{H1, H2, H3, H4, H5, H6}\}$$

Each value $v \in D(x)$ defines a possible value assignment $x = v$ to x. A set that contains exactly one of those value assignments for each variable in $\mathcal{X}$ and that contains no other elements is called an *assignment* to $\mathcal{X}$. For example, a wind-surfing vacation in Honolulu's H3 chain is represented by the assignment $\sigma_1 := \{x_1 = \text{WindSurfing}, x_2 = \text{Honolulu}, x_3 = \text{H3}\}$. The set of all assignments to $\mathcal{X}$ is called the *problem space* of $\mathcal{X}$. Given an assignment σ to $\mathcal{X}$ we can project it to a subset Y of the variables by choosing the value assignments to elements of Y:

$$\sigma[Y] := \{(x = v) \in \sigma \mid x \in Y\} \tag{1}$$

For example, projecting the assignment σ_1 to the vacation activity and the vacation destination results into $\sigma_1[\{x_1, x_2\}] = \{x_1 = \text{WindSurfing}, x_2 = \text{Honolulu}\}$.

2.2   Constraints

We can restrict the problem space of $\mathcal{X}$ by defining constraints on variables in $\mathcal{X}$. A constraint c has a scope $X_c \subseteq \mathcal{X}$ and a 'relation' which we express by a set R_c of assignments to the scope X_c. This set can be specified explicitly in form of a table where each column corresponds to a variable in X_c, each row corresponds to an assignment in R_c, and the value v from a value assignment $(x = v) \in \sigma$ is put in the cell for column x and row σ. Tables 1 and 2 show two constraints of the vacation example. The first constraint describes the activities that are possible in a city and has the scope $\{x_1, x_2\}$. The second constraint shows which hotel chain is available in which city and has the scope $\{x_2, x_3\}$. The relation R_c can also be specified by a logical formula that involves the variables from X_c and the operations from a given mathematical structure (such as arithmetic operations, boolean comparisons, and boolean operations).

A constraint satisfaction problem CSP for $\mathcal{X}$ is given by a finite set of constraints $\mathcal{C}$ the scopes of which are all subsets of $\mathcal{X}$. A CSP is finite if all its domains and relations are finite. A constraint c is satisfied by an assignment σ to $\mathcal{X}$ iff $\sigma[X_c]$ is an element of R_c. An assignment σ is a *solution* of $\mathcal{C}$ iff it satisfies all constraints of $\mathcal{C}$. If a CSP has no solution then it is called inconsistent. It is sometimes convenient to replace a CSP $\mathcal{C}$ by the conjunction $\bigwedge_{c \in \mathcal{C}} c$ of its constraints and vice versa.

Table 1. Constraint c_1

Activity x_1	City x_2
Casino	Antibes
Cliffdiving	Acapulco
FilmStudio	LosAngeles
SeaPark	Antibes
SeaPark	LosAngeles
WindSurfing	Antibes
WindSurfing	Honolulu
WindSurfing	Miami

Table 2. Constraint c_2

City x_2	Hotel chain x_3
Acapulco	H2
Acapulco	H6
Antibes	H3
Antibes	H5
Honolulu	H3
Honolulu	H5
LosAngeles	H2
LosAngeles	H4
LosAngeles	H6
Miami	H1
Miami	H4

2.3 Criteria and Preferences

A CSP can have multiple solutions and the user may prefer certain solutions to others. Since there may be an exponential number of solutions, it is not feasible to generate all solutions first and then to ask the user to express preferences between the solutions. The user will instead formulate preferences on certain properties of the solution such as the vacation region or the price of the vacation. These criteria are mathematical functions from the problem space to an outcome domain. Formally, a criterion z with domain Ω is an expression $f(x_1, \ldots, x_n)$ where $x_1, \ldots, x_n$ are variables from $\mathcal{X}$ and f is a function of signature $D(x_1) \times \ldots \times D(x_n) \to \Omega$. We suppose that the function f can be formulated with the operators of the constraint language (e.g. sum, min, max, conditional expression) or by a table. We can evaluate the expression $f(x_1, \ldots, x_n)$ if an assignment σ to the variables in $\mathcal{X}$ is given. We denote the resulting value by $z(\sigma)$.

For example, we want to express preferences on the activity, the vacation region, and the price and quality of the hotel chain. We introduce four criteria z_1, z_2, z_3, z_4 and their respective domains $\Omega_1, \Omega_2, \Omega_3, \Omega_4$. The criterion z_1 is equal to the vacation activity x_1 and has the domain $\Omega_1 := D(x_1)$. The other criteria are defined via a table (see Tables 3 and 4) and have the following domains:

$$\Omega_2 := \{\text{California, FrenchRiviera, Florida, Hawaii, Mexico}\}$$
$$\Omega_3 := [0, 1000]$$
$$\Omega_4 := \{\text{Economic, Basic, Standard, Luxury}\}$$

The user can compare the different outcomes in a domain Ω and formulate preferences between them. Preferences are modelled in form of a preorder $\succsim$ on Ω. The preorder consists of a strict part $\succ$ and an indifference relation $\sim$. If $\omega_1 \succsim \omega_2$ holds for two outcomes $\omega_1, \omega_2 \in \Omega$, then this means that the outcome ω_1 is at least as preferred as ω_2. We do not require that the preference order $\succsim$ is complete. If neither $\omega_1 \succsim \omega_2$, nor $\omega_2 \succsim \omega_1$ hold, then the preference between ω_1 and ω_2 is not specified and the user can refine it later on.

A preorder is a transitive and reflexive relation. It is not necessary to specify this order exhaustively. It can be obtained by determining the reflexive and

Table 3. Criterion for city	
City x_2	**Region** z_2
Acapulco	Mexico
Antibes	FrenchRiviera
Honolulu	Hawaii
LosAngeles	California
Miami	Florida

Table 4. Criteria for hotels		
Hotel x_3	**Price** z_3	**Quality** z_4
H1	40	Economic
H2	60	Basic
H3	100	Basic
H4	100	Standard
H5	150	Standard
H6	200	Luxury

transitive closure of a set $R \subseteq \Omega \times \Omega$ of user preferences. For example, suppose that the user prefers Hawaii at least as much as California, California at least as much as the French Riviera and as Florida, and the French Riviera at least as much as Mexico. Similarly, the user prefers cliff diving at least as much as sea park visits, sea park visits at least as much as casino visits and as wind-surfing, and casino visits and wind-surfing at least as much as visits to film studios. Finally, the user prefers casino visits at least as much as wind-surfing and vice versa. Figures 2 and 3 show these preferences (straight arcs and dotted-dashed arcs) and the corresponding preorders (any arc) in a graphical form.

We are mainly interested in the strict part $\succ$ of this preorder, namely the set of all pairs (ω_1, ω_2) in $\Omega \times \Omega$ such that $\omega_1 \succsim \omega_2$ holds, but not $\omega_2 \succsim \omega_1$. The absence of a strict preference between two outcomes can either signify indifference or incompleteness. The strict part of a preorder is a strict partial order, i.e. an irreflexive and transitive relation. We write $\omega_1 \succeq \omega_2$ as a short-hand for $\omega_1 \succ \omega_2$ or $\omega_1 = \omega_2$. The relation $\succeq$ is a subset of the preorder $\succsim$, but the inverse does not hold in general. In the example, the strict parts are obtained by suppressing the dotted-dashed arcs and the reflexive arcs of the form (ω, ω). We also consider the case where the preorder $\succsim$ is complete. The strict part of a complete preorder is a ranked order, i.e. a strict partial order satisfying the following property: if $\omega_1 \succ \omega_2$ then either $\omega_3 \succ \omega_2$ or $\omega_1 \succ \omega_3$. Ranked orders can be represented by utility functions u that map assignments to a numerical value such that $u(\omega_1) > u(\omega_2)$ iff $\omega_1 \succ \omega_2$. A complete preorder that is additionally anti-symmetric is a total order. The strict part of a total order is a strict total order, i.e. a strict partial order that satisfies $\omega_1 \succ \omega_2$ or $\omega_2 \succ \omega_1$ or $\omega_1 = \omega_2$.

The user can formulate preferences on multiple criteria. Consider m criteria $z_1, \ldots, z_m$ with domains $\Omega_1, \ldots, \Omega_m$. Furthermore, consider a strict partial order $\succ_i$ for each domain Ω_i. We say that the pair $p_i := \langle z_i, \succ_i \rangle$ of the i-th criterion and the i-th order is a *preference*. This terminology is, for example, used in [10]. We abbreviate a preference $\langle z_i, > \rangle$ using the increasing order $>$ by $maximize(z_i)$ and a preference $\langle z_i, < \rangle$ based on the decreasing order by $minimize(z_i)$. For example, price minimization is expressed as $minimize(z_3)$. Preferences thus generalize optimization objectives as used in CP or MIP.

2.4   Wishes

The user may also formulate wishes about the properties that a solution should have. For example, the user may wish to spend the vacation in Los Angeles.

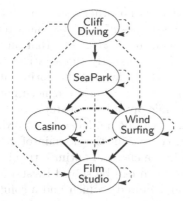

Fig. 2. Preferences $\succsim_1$ on activities

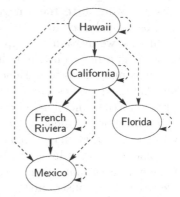

Fig. 3. Preferences $\succsim_2$ on regions

Such a wish is a soft constraint which should be satisfied if possible. Hence, a solution that satisfies a wish is preferred to a solution that violates the wish. A wish for constraint c can thus be modelled by a preference $\langle z_c, > \rangle$, which we abbreviate by $wish(c)$. It involves a binary criterion, namely the truth value of the constraint, and an implicit preference ordering that prefers true to false. Given an assignment σ, the truth value z_c of c is defined as follows:

$$z_c(\sigma) := \begin{cases} 1 \text{ if } \sigma \text{ satisfies } c \\ 0 \text{ otherwise.} \end{cases} \qquad (2)$$

3   Preference-Based Problem Solving

Combinatorial problems with preferences are classically solved by compiling all preferences into a single utility function and by determining a solution of the constraints that has maximal or nearly maximal utility. However, the optimizer does not give an explanation of optimality in terms of the original preferences. If the user specified a preference $\langle z, \succ \rangle$, she wants to know whether the criterion z has its best value in the solution. If not, the user wants to get an explanation why no better outcome has been obtained for the criterion. If the criterion z is in conflict with other criteria then the explanation should indicate this conflict and the trade-off that has been made. We show that explanations of optimality can be produced if the problem solving process is based on the original preferences. The form of the explanation will depend on the kind of the optimization problem. We consider single-criterion optimization, different variants of lexicographic optimization, and Pareto-optimization.

3.1   Atomic Optimization Step

Given a single preference $\langle z, \succ \rangle$, we are interested in those solutions of the constraints C that assign a $\succ$-maximal value to the criterion z. A solution σ assigns a $\succ$-maximal value v to the criterion z iff there is no other solution σ^* that assigns a better value v^* to z, i.e. a value that satisfies $v^* \succ v$. To characterize those solutions, we introduce an operator, denoted by $Max(\langle z, \succ \rangle)$, that maps a constraint C to a new constraint that is satisfied by exactly the solutions that assign $\succ$-maximal values to z. Hence, $Max(\langle z, \succ \rangle)(C)$ denotes the optimization problem that need to be solved. As it concerns a single criterion, it need not be decomposed further and thus represents an atomic optimization step.

If $>$ is a total order, then the preference $\langle z, > \rangle$ can be modelled by an ordinal utility function u that maps each possible outcome ω in the domain of z to a unique numeric utility value $u(\omega)$. We then obtain a classical optimization problem, namely that of maximizing $u(z)$. This problem can be solved by standard optimizers (such as constraint-based Branch-and-Bound), which find a solution of maximum value u^* for $u(z)$. Since we supposed that $>$ is a total order, there is a unique outcome ω^* in the domain of z that has the utility u^*. Hence, each solution of C that is optimal w.r.t. the preference $\langle z, > \rangle$ satisfies the constraint $C \wedge z = \omega^*$ and vice versa. Hence, the following equivalence holds for total orders:

$$Max(\langle z, > \rangle)(C) \equiv C \wedge z = \omega^* \tag{3}$$

We can thus characterize the entire set of optimal solutions by the constraint $C \wedge z = \omega^*$ and replace the original problem $Max(\langle z, > \rangle)(C)$ by this constraint without loosing any optimal solution. Subsequent optimization steps can then further reduce the set of optimal solutions of $Max(\langle z, > \rangle)(C)$. We also say that $C \wedge z = \omega^*$ is the *solved form* of the optimization problem $Max(\langle z, > \rangle)(C)$.

Once the optimization problem has been solved, the user may ask for explanations of optimality such as

1. Why can't z have a value better than ω^*?
2. Why hasn't the value ω been chosen for z?

As there is no better value than ω^* for z, the conjunction of C and the unary constraint $z > \omega^*$ is inconsistent. As the constraint $z > \omega^*$ is unary, it can easily be encoded in a constraint solver (e.g. by removing all values smaller than or equal to ω^* from the domain of z). We can now determine an explanation by asking why the set C' of the conjuncts of $C \wedge z > \omega^*$ is inconsistent. To answer this question, we determine a conflict for C', i.e. a minimal subset X' of C' that is inconsistent. This conflict can, for example, be computed by the QUICKXPLAIN-algorithm [6]. Since C is assumed to be consistent, the conflict X' needs to contain the constraint $z > \omega^*$. The other elements $X := X' \setminus \{z > \omega^*\}$ of the conflict then explain why z can't have a value better than ω^* w.r.t. the order $>$. These constraints X defeat any value for z that is better than the optimum ω^*. The user might also ask why z has not obtained a specific value ω. If ω is greater than the optimum ω^*, then the choice $z = \omega$ is defeated by X. Otherwise, ω is smaller than the optimum ω^* and has not been chosen for that reason. In order to give the right answer, the explanation of optimality needs to include the defeaters X and the ordering $>$.

Let σ be a solution of the problem $Max(q)(C)$ with total preferences $q := \langle z, > \rangle$. An *explanation of the Max(q)-optimality* of σ is a triple (q, ω^*, X) such

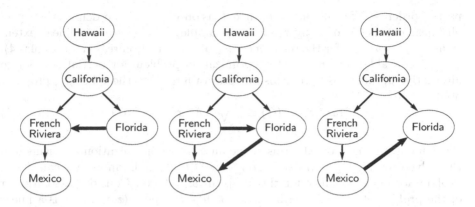

Fig. 4. The three linear extensions $>_{2,1}, >_{2,2}, >_{2,3}$ of the region preferences

that ω^* is equal to the optimum $z(\sigma)$ and X is a minimal subset of the set of conjuncts of C for which $X \cup \{z > \omega^*\}$ is inconsistent.

An example is that of minimizing the hotel price in the vacation example. The smallest value in the price domain is 0, but obviously this value is not possible as image of z_3. Indeed, the optimal price of the problem $Max(minimize(price))(C)$ is 40 and an explanation of optimality is $(minimize(price), 40, \{\})$.

We are thus able to provide explanations of optimality for atomic optimization problems with total preferences.

3.2   Alternative Optimizations

In the beginning of an interactive problem solving process, the user will usually specify only a partial order. Hence, there may be two solutions σ_1 and σ_2 of $Max(\langle z, \succ \rangle)$ that assign two different values ω_1 and ω_2 to the criterion z such that neither ω_1 is strictly preferred to ω_2, nor ω_2 is strictly preferred to ω_1. As a consequence, the solved form of $Max(\langle z, \succ \rangle)(C)$ will be a disjunction of assignments of the form $z = \omega_i$. If Ω^* is the set of all optimal outcomes, i.e. the values that are assigned to the criterion z in the different solutions of the optimization problem, then the solved form of the problem is as follows:

$$Max(\langle z, \succ \rangle)(C) \equiv \bigvee_{\omega^* \in \Omega^*} (C \wedge z = \omega^*) \qquad (4)$$

An example is the partial preference order $\succ_2$ on the vacation regions. Suppose that the cities Honolulu and Los Angeles are not possible due to booking problems. The problem $Max(\langle z_2, \succ_2 \rangle)(C')$ where $C' := C \wedge x_2 \neq$ Honolulu $\wedge\, x_2 \neq$ LosAngeles has the optimal outcomes Florida and French Riviera. The solved form is $C' \wedge (z_2 = $ Florida $\vee\, z_2 = $ FrenchRiviera$)$.

We can compute this solved form by reducing the optimization problem with partial orders to multiple optimization problems with total orders. For this purpose, we consider a strict total order $>$ on Ω that is a superset of the strict partial order $\succ$. We call this a linear extension of $\succ$ and we denote the set of all linear extensions of $\succ$ by $\tau(\succ)$. For example, we obtain three linear extensions $>_{2,1}, >_{2,2}, >_{2,3}$ for the preferences $\succ_2$ on the vacation region (cf. Figure 4). We are now able to transform the optimization problem for partial orders into alternative optimization problems on total orders since the following property holds:

$$Max(\langle z, \succ \rangle)(C) \equiv \bigvee_{>\in\tau(\succ)} Max(\langle z, > \rangle)(C) \qquad (5)$$

We can combine the solved forms of the alternative optimization problems into the solved form for the problem $Max(\langle z, \succ \rangle)(C)$. Furthermore, we can define explanations of optimality for this problem as follows. Consider a solution σ of the problem $Max(p)(C)$ with partial preferences $p := \langle z, \succ \rangle$. If $>$ is a linear extension of $\succ$, $q := \langle z, > \rangle$ and σ is a solution of $Max(q)(C)$, then an explanation of the $Max(q)$-optimality of σ is also an *explanation of the $Max(p)$-optimality* of σ. Due to (5), such a linear extension $>$ exists for each solution σ of $Max(p)(C)$.

In the example $Max(\langle z_2, \succ_2 \rangle)(C')$, the linear extension $>_{2,1}$ prefers Florida to the French Riviera and thus allows us to explain the optimality of Florida. The other two linear extensions, namely $>_{2,2}, >_{2,3}$, prefer French Riviera to Florida. Any of them can be used in an explanation of the optimality of the French Riviera. An example is:

$$\xi := (\langle z_2, >_{2,2} \rangle, \text{FrenchRiviera}, \{x_2 \neq \text{Honolulu}, x_2 \neq \text{LosAngeles}\})$$

Given this explanation, the user can ask why the other regions have not been selected. The options Hawaii and California are defeated by the constraints $\{x_2 \neq$ Honolulu, $x_2 \neq$ LosAngeles$\}$. The option Mexico has been discarded since the chosen option French Riviera is preferred to Mexico. And the option Florida has been discarded since the linear extension $>_{2,2}$ prefers the French Riviera to Florida. The user can criticize this response by adding the preference Florida $\succ'_2$ FrenchRiviera. This will eliminate the solutions for the French Riviera.

The linear extensions of the user preferences are an important part of the explanation as they give hints to the user how to eliminate undesired solutions. As several linear extensions may support the same solution, it is not necessary to enumerate a factorial number of linear extensions when computing the set of all optimal outcomes [7].

3.3   Lexicographical Approach

User preferences do not concern a single criterion, but multiple criteria. We therefore consider the preferences $p_1, \ldots, p_n$ on n criteria. We suppose that p_i has the form $\langle z_i, \succ_i \rangle$. Each preference gives rise to an optimization operator $Max(p_i)$. If the conjunction $\bigwedge_i Max(p_i)(C)$ has a solution, then the preferences are not in conflict and we obtain ideal solutions. Otherwise, we need to find a trade-off between the preferences. The easiest way is to introduce an importance order on the preferences and to decide trade-offs in favour of the more important preferences. Lexicographic optimization defines an ordering on the solution space based on this importance principle. Consider two assignments σ_1 and σ_2. We consider the tuples of values that both assignments assign to the criteria and define a lexicographical order $\succ_{lex}$ between them. The relation

$$(z_1(\sigma_1), \ldots, z_n(\sigma_1)) \succ_{lex} (z_1(\sigma_2), \ldots, z_n(\sigma_2)) \tag{6}$$

holds iff there exists a k such that $z_k(\sigma_1) \succ_k z_k(\sigma_2)$ and $z_i(\sigma_1) = z_i(\sigma_2)$ for $i = 1, \ldots, k - 1$. A solution σ of C is a *lexicographically optimal solution* of C iff there is no other solution σ^* of C such that $(z_1(\sigma^*), \ldots, z_n(\sigma^*))$ is lexicographically better than $(z_1(\sigma), \ldots, z_n(\sigma))$. We introduce the lexicographic optimization operator $Lex(p_1, \ldots, p_n)$ that maps a constraint C to a new constraint C' that is satisfied by the lexicographically optimal solutions of C.

Similar to (5), we can transform the problem $Lex(p_1, \ldots, p_n)(C)$ into a solved form by considering a linear extension for each strict partial order $\succ_i$. We define $\tau(\langle z_i, \succ_i \rangle)$ as the set of all preferences $\langle z_i, >_i \rangle$ for which $>_i$ is a linear extension of $\succ_i$. Furthermore, we define $\tau(p_1, \ldots, p_n)$ as the Cartesian product $\tau(p_1) \times \ldots \times \tau(p_n)$. The following equivalence holds between lexicographic optimization problems with partial preferences and the problems obtained by linearizing these preferences:

$$Lex(p_1, \ldots, p_n)(C) \equiv \bigvee_{(q_1, \ldots, q_n) \in \tau(p_1, \ldots, p_n)} Lex(q_1, \ldots, q_n)(C) \tag{7}$$

A lexicographic optimization problem $Lex(q_1, \ldots, q_n)(C)$ with total preferences can then be transformed to a sequence of single-criterion optimization problems which can be solved by a standard optimizer:

$$\begin{aligned} Lex(q_1)(C) &\equiv Max(q_1)(C) \\ Lex(q_1, \ldots, q_n)(C) &\equiv Lex(q_2, \ldots, q_n)(Max(q_1)(C)) \end{aligned} \tag{8}$$

The solved form has the form $C \wedge z = \omega_1^* \wedge \ldots z = \omega_n^*$. In the vacation example, the problem $Lex(\langle z_1, \succ_1 \rangle, \langle z_2, \succ_2 \rangle, minimize(z_3))(C)$ has the solved form $C \wedge z_1 = $ CliffDiving $\wedge z_2 = $ Mexico $\wedge z_3 = 60$.

Explanations for lexicographical optimality are sequences of explanations for single-criterion optimization problems. Consider a solution σ of the problem $Lex(p_1, \ldots, p_n)(C)$. A sequence $(\xi_1, \ldots, \xi_n)$ is called an *explanation of the $Lex(p_1, \ldots, p_n)$-optimality* of σ iff there exist totally ordered preferences $(q_1, \ldots, q_n) \in \tau(p_1, \ldots, p_n)$ such that ξ_i is an explanation of $Max(q_i)$-optimality of the i-th optimization problem $Max(q_i)(C \wedge z_1 = z_1(\sigma) \wedge \ldots \wedge z_{i-1} = z_{i-1}(\sigma))$ and σ is a solution of this problem. Explanations of $Lex(p_1, \ldots, p_n)$-optimality always exist and can easily be produced when solving the problem.

Consider a solution σ_1 of the vacation configuration problem $Lex(\langle z_1, \succ_1 \rangle, \langle z_2, \succ_2 \rangle, minimize(z_3))(C)$. An explanation of optimality is (ξ_1, ξ_2, ξ_3) where

$$\begin{aligned} \xi_1 &:= (\langle z_1, >_{1,1} \rangle, \text{CliffDiving}, \{\}) \\ \xi_2 &:= (\langle z_2, >_{2,1} \rangle, \text{Mexico}, \{c_1, z_1 = \text{CliffDiving}\}) \\ \xi_3 &:= (\langle z_3, < \rangle, 60, \{c_2, z_2 = \text{Mexico}\}) \end{aligned}$$

We can depict this explanation in a graphical form. Each triple ξ_i is represented by a node. There is an edge from ξ_i to ξ_j if the defeaters of z_j contain a constraint of the form $z_i = \omega_i$ meaning that the optimal value for z_i helped to defeat the better values for z_j. Figure 7 shows the explanation of the lexicographical optimality of σ_1 in graphical form. The edge between the nodes ξ_1 and ξ_2 indicates that the two criteria z_1 and z_2 are in conflict and that the conflict has been resolved in favour of the more important criteria, namely the activity z_1.

3.4   Alternative Sequentializations

Whereas lexicographical optimization is one of the fundamental approaches of multi-criteria optimization, few attention has been paid to the choice of the importance ordering. In the vacation example, is it more important to minimize the price than to maximize the quality? Or is it more important to maximize the quality than to minimize the price? It may be worth to compute both of those 'extreme solutions' before before exploring compromises. This is reasonable if there is a small number of criteria having the same importance.

We characterize each family of extreme solutions by a permutation π of the n preferences $(p_1, \ldots, p_n)$. Given such a permutation, we optimize the criteria in the ordering $p_{\pi_1}, \ldots, p_{\pi_n}$. Let Π be the set of all those permutations. We introduce a new operator, called $Permute(p_1, \ldots, p_n)(C)$, that maps the constraint C to a constraint C' that is satisfied by all extreme solutions of C. This constraint is equivalent to a disjunction of lexicographic optimization problems:

$$Permute(p_1, \ldots, p_n)(C) \equiv \bigvee_{\pi \in \Pi} Lex(p_{\pi_1}, \ldots, p_{\pi_n})(C) \qquad (9)$$

As explanations of the different lexicographic optimization problems preserve the ordering of the criteria, it is straightforward to combine the explanations of those problems. Consider a solution σ of the problem $Permute(p_1, \ldots, p_n)(C)$. If π is a permutation such that σ is solution of $Lex(p_{\pi_1}, \ldots, p_{\pi_n})(C)$, then an explanation $(\xi_{\pi_1}, \ldots, \xi_{\pi_n})$ of the $Lex(p_{\pi_1}, \ldots, p_{\pi_n})$-optimality of σ is an *explanation of the $Permute(p_1, \ldots, p_n)$-optimality* of σ.

The vacation example $Permute((z_1, \succ_1), (z_2, \succ_2), minimize(z_3))(C)$ has three extreme solutions that are justified by different importance orderings. Solution σ_1 chooses cliff diving in Mexico with hotel costs of \$60 based on the order z_1, z_2, z_3. Solution σ_2 proposes wind-surfing in Hawaii with hotel costs of \$100 by following the order z_2, z_1, z_3. Solution σ_3 offers wind-surfing in Florida with hotel costs of \$40 based on the order z_3, z_2, z_1. Figure 5 displays the possible combinations of the criteria values (in a way that is similar to the micro-structure of constraint networks). The three extreme solutions are represented by thick lines. As an example of an explanation, we give one for the optimality of the last solution:

$$(((z_3, <), 40, \{\}),$$
$$((z_2, >_{2,1}), Florida, \{c_2, z_3 = 40\}),$$
$$((z_1, >_{1,1}), WindSurfing, \{c_1, z_2 = Florida\}))$$

This explanation exhibits the importance ordering of the criteria and thus explains how conflicts between preferences are resolved.

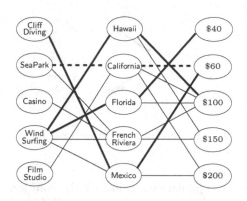

Fig. 5. Combinations of criteria

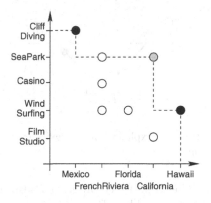

Fig. 6. Trade-off between z_1 and z_2

3.5   Importance Preferences

Thanks to the explanations, the user can inspect the conflicts between criteria and the way they have been resolved. If the user is not satisfied with such a conflict resolution, she can change it by reordering the criteria. For example, suppose that the user is not satisfied with a vacation package at a high price. The user now learns that this high price is caused by a very good quality rating, which was chosen to be more important. The user wants to give price minimization a higher importance than quality maximization and thus expresses an importance ordering between the price criterion z_3 and the quality criterion z_4. We formalize this importance ordering in terms of a strict partial order $I \subseteq \mathcal{Z} \times \mathcal{Z}$ on the criteria set $\mathcal{Z} := \{z_1, \ldots, z_n\}$. In our example, we have

$$I := \{(z_3, z_4)\} \tag{10}$$

We now use this importance ordering to restrict the set of extreme solutions. We consider only those permutations π of the preferences $p_1, \ldots, p_n$ that respect the importance ordering. More important criteria need to be ranked first. Hence, the permutation π respects the importance ordering I iff the following property holds for all i, j:

$$(z_{\pi_i}, z_{\pi_j}) \in I \text{ implies } i < j \tag{11}$$

Let $\Pi(I)$ be the set of all permutations π that respect I. Furthermore, we introduce a variant of the permute-operator $Permute(p_1, \ldots, p_n : I)(C)$ which is restricted to those extreme solutions that are obtained by the permutations in $\Pi(I)$:

$$Permute(p_1, \ldots, p_n : I)(C) \equiv \bigvee_{\pi \in \Pi(I)} Lex(p_{\pi_1}, \ldots, p_{\pi_n})(C) \tag{12}$$

Preferences between criteria may eliminate extreme solutions, but do not add new ones:

$$I \subseteq I^* \text{ implies } Permute(p_1, \ldots, p_n : I^*)(C) \Rightarrow Permute(p_1, \ldots, p_n : I)(C) \tag{13}$$

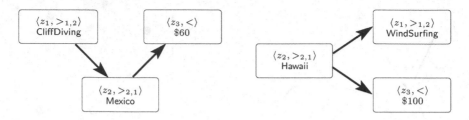

Fig. 7. Explanation of lexicographic optimality of σ_1

Fig. 8. Explanation of lexicographic optimality of σ_2

The importance ordering I also impacts explanations. Consider the solution σ of the problem $Permute(p_1, \ldots, p_n : I)(C)$. If π is a permutation in $\Pi(I)$ and σ is a solution of $Lex(p_{\pi_1}, \ldots, p_{\pi_n})(C)$, then an explanation $(\xi_{\pi_1}, \ldots, \xi_{\pi_n})$ of the $Lex(p_{\pi_1}, \ldots, p_{\pi_n})$-optimality of σ is an *explanation of the Permute*$(p_1, \ldots, p_n : I)$-*optimality* of σ.

We now discuss the effect of importance preferences on the vacation example. Suppose that solution σ_3 has been submitted to the user. This solution proposes wind-surfing in Florida with hotel costs of \$40 based on the ordering z_3, z_2, z_1. The user criticizes this explanation by stating that the choices of the vacation activity z_1 and of the vacation region z_2 are more important than the price z_3. The importance preferences $I_1 := \{(z_1, z_3), (z_2, z_3)\}$ are added to the preference model. As the order z_3, z_2, z_1 does not respect these importance preferences, solution σ_3 is no longer proposed. Solutions σ_1 and σ_2 have importance orderings that represct I_1. Figures 7 and 8 give explanations of optimality of these teo solutions of the problem $Permute(\langle z_1, \succ_1 \rangle, \langle z_2, \succ_2 \rangle, minimize(z_3) : I_1)(C)$.

3.6   Trade-Offs and Preference Limits

Extreme solutions are resolving conflicts between preferences completely in favour of the more important criteria. If two criteria z_1 and z_2 are in conflict, then z_1 gets its best value, while z_2 is completely penalized. Changing the order completely turns the balance around and the conflict is decided in favour of z_2 while penalizing z_1. However, it is also possible to use compromises and to trade a small improvement for z_2 against a small penalization of z_1 without changing the order of the criteria. Pareto-optimality captures all the possible trade-offs.

The notion of Pareto-optimality does not impose any importance ordering on the criteria. It extends the partial ordering on the different criteria to a partial ordering on the complete criteria space without making any particular assumption. An assignment σ^* *Pareto-dominates* another assignment σ iff 1. σ^* and σ differ on at least one criterion z_k (i.e. $z_k(\sigma^*) \neq z_k(\sigma)$) and 2. σ^* is strictly better than σ on all criteria on which they differ (i.e. $z_i(\sigma^*) \neq z_i(\sigma)$ implies $z_k(\sigma^*) \succ_k z_k(\sigma)$). An equivalent definition consists in saying that σ^* dominates σ iff σ^* is at least as as good as σ on all criteria (i.e. $z_k(\sigma^*) \succeq_k z_k(\sigma)$) and there is at least one criterion where σ^* is strictly better than σ (i.e. $z_k(\sigma^*) \succ_k z_k(\sigma)$). Pareto-dominance defines a strict partial order on the criteria space.

A solution σ of C is a *Pareto-optimal solution* of C iff there is no other solution σ^* that Pareto-dominates σ. We introduce an operator $Pareto(p_1, \ldots, p_n)$ that maps a constraint C to a new constraint C' that is only satisfied by the Pareto-optimal solutions of C. Non-Pareto-optimal solutions clearly are not desirable since there are solutions that are better on one or more criteria while keeping the other criteria unchanged.

Figure 5 shows a Pareto-optimal solution for the vacation example which is not an extreme solution. This solution, which we name σ_4, proposes sea park visits in California with hotel costs of \$60. Hence, σ_4 does not choose the best value for any criteria, but it is Pareto-optimal since we cannot improve any criterion without getting a worse value for another criterion. Figure 6 shows the trade-off between the vacation activity z_1 and the vacation region z_2. The solution σ_4 is situated between the two extreme solutions σ_1 and σ_2.

As for lexicographical optimization, we can linearize the partially ordered preferences and transform a Pareto-optimization problem into a disjunction of Pareto-optimization problems with totally ordered preferences:

$$Pareto(p_1, \ldots, p_n)(C) \equiv \bigvee_{(q_1, \ldots, q_n) \in \tau(p_1, \ldots, p_n)} Pareto(q_1, \ldots, q_n)(C) \qquad (14)$$

However, there is no direct way to transform a Pareto-optimization problem into a solved form even if it is based on totally ordered preferences. One approach to solve those problems consists in generalizing optimization methods such as Branch-and-Bound search to a partial order. The approach is pursued in [4]. Branch-and-bound search for a partial order needs to maintain a whole Pareto-optimal frontier which might become rather inefficient. More importantly, this method does not produce explanations of optimality that exhibit the trade-offs between criteria. For this reason, we do not follow this approach, but seek ways to solve Pareto-optimization problems by alternative sequences of classical optimization steps. We observe that all extreme solutions are Pareto-optimal (see [5]):

$$Permute(p_1, \ldots, p_n)(C) \Rightarrow Pareto(p_1, \ldots, p_n)(C) \qquad (15)$$

Hence, we can start with extreme solutions when determining Pareto-optimal solutions. An extreme solution is entirely characterized by an ordering of the criteria (and of the user preferences). However, these orderings do not characterize those compromises between two conflicting criteria where none of the two criteria gets its best value. We need to insert additional steps into the sequence of optimization problems. An extreme solution is always in favour for the most important criterion and completely penalizes the less important ones. For example, consider two conflicting preferences $\langle z_1, \succ_1 \rangle$ and $\langle z_2, \succ_2 \rangle$. Let σ be the extreme solution for the ordering z_1, z_2. The criterion z_1 gets its best feasible value, namely $z_1(\sigma)$, whereas z_2 is penalized and gets the value $z_2(\sigma)$. If we want to obtain a better value for the less important criterion z_2, then we need to limit its penalization before optimizing the more important criterion z_1. For this purpose, we add the constraint $z_2 \succ z_2(\sigma)$ before optimizing z_1. If this constraint is satisfiable, then the optimization of z_1 will produce the best solution for z_1

under the constraint $z_2 \succ z_2(\sigma)$. If such a penalization limit is infeasible, then it should be retracted. To achieve this, we represent penalization limits of the form $z_2 \succ z_2(\sigma)$ as wishes.

We introduce a wish for each criterion and for each possible value of this criterion:

$$limits(\langle z_i, \succ_i \rangle) := \langle wish(z_i \succeq_i \omega_1), \ldots, wish(z_i \succeq_i \omega_n) \rangle \qquad (16)$$

Furthermore, we consider the set I of importance preferences stating that wishes for worse outcomes precede wishes for better outcomes:

$$I := \{(wish(z_i \succeq_i \omega), wish(z_i \succeq_i \omega^*)) \mid \omega^* \succ \omega, i = 1, \ldots, n\} \qquad (17)$$

It is a common modelling technique in Operations Research to transform an integer variable into a set of binary variables. Hence, our transformation has nothing unusual except that it is applied to the criteria and not to the decision variables. The Pareto-optimal solutions of the original model then correspond exactly to the extreme solutions of the binary model (cf. theorem 1 in [5]):

$$Pareto(p_1, \ldots, p_n)(C) \equiv Permute(limits(p_1), \ldots, limits(p_n) : I)(C) \qquad (18)$$

This correspondence allows us to transform Pareto-optimal solutions into a solved form and to define explanations of optimality.

Interestingly, the original optimization steps of the form $Max(\langle z_i \succ_i \rangle)(C')$ have disappeared in this new characterization. If the result of this optimization is ω', then this step corresponds to the last successful wish on z_i, namely $Max(wish(z_i \succeq_i \omega'))(C')$. Each Pareto-optimal solution is thus characterized by a sequence of wishes and there are multiple wishes for the same criterion z_i. There are logical dependencies between wishes that allow us to speed up the solving process and to simplify the explanations. If $wish(z_i \succeq_i \omega)$ fails, then all wishes $wish(z_i \succeq_i \omega')$ for better outcomes $\omega' \succ \omega$ will also fail. Furthermore, if $wish(z_i \succeq_i \omega)$ succeeds and is directly preceded by a wish for the same criterion then it subsumes this previous wish and the previous wish can be removed from explanations of optimality.

Explanations of Pareto-optimality are thus obtained by determining subsequences of explanations for lexicographical optimality of the binary preference model. Let σ be a Pareto-optimal solution of $Pareto(p_1, \ldots, p_n)(C)$. Then σ is an extreme solution of $Permute(limits(p_1), \ldots, limits(p_n) : I)(C)$ and there exists an explanation $(\xi_1, \ldots, \xi_m)$ of optimality of this extreme solution. Let ξ_j be $(\langle wish(z'_j \succeq \omega'_j), \succ \rangle, v_j, X_j)$. We say that ξ_j is relevant iff the wish is successful (i.e. $v_j = 1$) and the next triple concerns a different criterion (i.e. $z'_{j+1} \neq z'_j$). An explanation of Pareto-optimality of σ is the sequence $(\xi_{j_1}, \ldots, \xi_{j_k})$ of the relevant triples from $(\xi_1, \ldots, \xi_n)$ in the original order, i.e. $j_1 < j_2 < \ldots < j_k$. These explanations can still contain multiple wishes for the same criterion. The last wish for the criterion in an explanation determines the value of the criterion. The other wishes limit the penalization of the criterion before optimizing other criteria. Hence, the explanation highlights the critical choices that need to be made in order to obtain the Pareto-optimal solution σ.

The solving algorithm for Pareto-optimal solutions in [5] is based on wishes and is easily able to provide these explanations of optimality.

We give an explanation for the Pareto-optimal solutions σ_1 and σ_4 of the vacation example. The explanation for σ_1 consists of one wish for each criterion. Each of these wishes assigns the optimal value.

$$((wish(z_1 \succeq_1 \text{CliffDiving}), 1, \{\}),$$
$$(wish(z_2 \succeq_2 \text{Mexico}), 1, \{c_1, z_1 \succeq_1 \text{CliffDiving}\}),$$
$$(wish(z_3 \leq 60), 1, \{c_2, z_2 \succeq_2 \text{Mexico}\}))$$

When this explanation is presented to the user, she might criticize it by saying that the vacation region has been penalized too much by its defeater, which is the wish $w_1 := wish(z_1 \succeq_1 \text{Cliffdiving})$. The user therefore adds a wish $w_2 := wish(z_2 \succeq_2 \text{FrenchRiviera})$ to limit this penalization. This wish needs to get higher importance than w_1 to be effective in all cases. The user therefore adds the importance statement (w_2, w_1) as well. This leads to a new solution, namely σ_4 and the following explanation:

$$((wish(z_2 \succeq_2 \text{FrenchRiviera}), 1, \{\}),$$
$$(wish(z_1 \succeq_1 \text{SeaPark}), 1, \{c_1, z_2 \succeq_2 \text{FrenchRiviera}\}),$$
$$(wish(z_2 \succeq_2 \text{California}), 1, \{c_1, z_1 \succeq_1 \text{SeaPark}\}),$$
$$(wish(z_3 \leq 60), 1, \{c_2, z_2 \succeq_2 \text{California}\}))$$

Explanations for Pareto-optimal, which consist of sequences of successful wishes, thus offer new ways to the user to explore the space of Pareto-optimal solutions.

4   Conclusion

We have shown that combinatorial optimization can directly use the original user preferences even if those preferences are incomplete. The solving process considers different ways to complete these preferences and optimizes a single criterion at a time, while exploring different importance orderings of the criteria. In doing so, the whole optimization process not only results in an optimal solution, but also produces an explanation of optimality of the solution. Such an explanation indicates the conflicts between preferences and shows how they have been resolved. The user can examine this explanation and either accept the solution or refine the preference model. The refined preferences will eliminate the undesired explanation. The problem solver may then find another explanation for the same solution or another solution and the procedure is repeated.

As explanations are comprehensible and are formulated in the same "language" as the optimization problem, the user can react to all elements of the explanation and change the problem statement. For example, the user can relax constraints, refine preferences, add importance statements between preferences, or limit the penalization of the less important criteria. Thanks to the explanations, the problem solver behaviour becomes completely transparent to the user and the user gains full control over the problem solver. We thus obtain an interactive problem solving process that consists of optimization, explanation, and preference elicitation. This offers new possibilities over a traditional MAUT approach which aggregates all preferences into a utility function. The MAUT approach is convenient for MIP optimizers, but not for producing explanations of optimality in terms of the original preferences, which help to make solutions understandable for the end users.

The approach can be extended to conditional preferences [1]. Algorithms for computing the optimal solutions can be found in [5,7].

References

1. Boutilier, C., Brafman, R.I., Domshlak, C., Hoos, H.H., Poole, D.: Preference-based constrained optimization with CP-nets. Computational Intelligence 20, 137–157 (2004)
2. Doyle, J.: Prospects for preferences. Computational Intelligence 20, 11–136 (2004)
3. Ehrgott, M.: A characterization of lexicographic max-ordering solutions. In: Methods of Multicriteria Decision Theory: Proceedings of the 6th Workshop of the DGOR Working-Group Multicriteria Optimization and Decision Theory, Egelsbach, pp. 193–202. Häsel-Hohenhausen (1997)
4. Gavanelli, M.: An algorithm for multi-criteria optimization in CSPs. In: ECAI 2002, pp. 136–140 (2002)
5. Junker, U.: Preference-based search and multi-criteria optimization. Annals of Operations Research 130, 75–115 (2004)
6. Junker, U.: QuickXplain: Preferred explanations and relaxations for over-constrained problems. In: AAAI 2004, pp. 167–172 (2004)
7. Junker, U.: Outer branching: How to optimize under partial orders? In: ECAI 2006 Workshop on Advances in Preference Handling, pp. 58–64 (2006)
8. Junker, U., Mailharro, D.: Preference programming: Advanced problem solving for configuration. AI-EDAM 17(1) (2003)
9. Keeney, R.L., Raiffa, H.: Decisions with Multiple Objectives. Wiley, Chichester (1976)
10. Kießling, W.: Foundations of preferences in database systems. In: 28th International Conference on Very Large Data Bases (VLDB 2002), pp. 311–322 (2002)

Generalizing Global Constraints Based on Network Flows

Igor Razgon[1,2], Barry O'Sullivan[1,2], and Gregory Provan[2]

[1] Cork Constraint Computation Centre, University College Cork, Ireland
[2] Department of Computer Science, University College Cork, Ireland
{i.razgon,b.osullivan,g.provan}@cs.ucc.ie

Abstract. Global constraints are used in constraint programming to help users specify patterns that occur frequently in the real world. In addition, global constraints facilitate the use of efficient constraint propagation algorithms for problem solving. Many of the most common global constraints used in constraint programming use filtering algorithms based on network flow theory. We show how we can formulate global constraints such as GCC, Among, and their combinations, in terms of a tractable set-intersection problem called Two Families Of Sets (TFOS). We demonstrate that the TFOS problem allows us to represent tasks that are often difficult to model in terms of a classical constraint satisfaction paradigm. In the final part of the paper we specify some tractable and intractable extensions of the TFOS problem. The contribution of this paper is the characterisation of a general framework that helps us to study the tractability of global constraints that rely on filtering algorithms based on network flow theory.

1 Introduction

Global constraints are used in constraint programming to help users specify patterns that occur frequently in the real world (see, for example, [8,9,10,11,12,13]). In addition, global constraints facilitate the use of efficient constraint propagation algorithms for problem-solving. Many of the most common global constraints used in constraint programming employ filtering algorithms based on network flow theory. Examples are the Global Cardinality Constraint (GCC) [11], and the Among Constraint [3], which generalize a number of other global constraints such as NotAllEqual, Max, and Member constraints.

A real-life problem usually needs *combinations* of global constraints, rather than a single constraint. Some of these combinations may be efficiently solved, but generally they are intractable [13]. A question that arises in this context is to describe a tractable problem that can represent various combinations of network flow-based global constraints. This question is addressed in this paper. In particular, our contributions are as follows.

1. We define a model that includes a ground set V and two families F_1 and F_2 of subsets of V. Any two elements of each family are either disjoint or contained one in the other. Each subset of V contained in these families is associated with two nonnegative integers called minimal and maximal cardinalities. We refer to the model as TFOS, which is an acronym for Two Families Of Sets.

F. Fages, F. Rossi, and S. Soliman (Eds.): CSCLP 2007, LNAI 5129, pp. 127–141, 2008.

2. Given a TFOS model (V, F_1, F_2) we say that a subset of V is valid if the size of its intersection with each set contained in F_1 or F_2 lies between the cardinalities assigned to that set. We define a TFOS problem as finding the largest valid subset of V and study the tractability of the problem.
3. We show that GCC and Among constraints, as well as (some of) their combinations considered in [13], can be represented by the TFOS model. In the proposed representation, V is the set of all values of the CSP being considered, F_1 is the family of all domains, F_2 represents the global constraints. By introducing additional sets to F_1, we demonstrate that it is possible to represent as a TFOS problem some optimization tasks that seem difficult to express as a classical CSP.
4. We propose a propagation algorithm that can be used to speed up search in cases where part of some intractable problem is presented as a TFOS problem. The propagation algorithm is based on the approach suggested in [11].
5. We discuss possible extensions of the TFOS model. In particular we show tractability of the weighted TFOS problem. Then we prove that introducing an additional family of sets of V with the same properties as the families of the TFOS model makes the resulting problem NP-hard. Finally, we show that we can preserve polynomial solvability by restricting the properties of the third family.

Several other approaches to the design of generalized global constraints are described in [2,4,13]. The work reported in [13] is most closely related to our approach: it proves the tractability of two types of combinations of GCC and Among constraints. However, the TFOS model is more general because, as we show further in this paper, it can express the combinations of constraints considered in [13] as well as a number of additional combinations of constraints that seem hard to express by that approach. The authors of [2] propose a method to design tractable logical combinations of some primitive constraints. However, their method is unable to express some basic network flow-based global constraints such as Among. The approach described in [4] has an emphasis different from ours: it aims at the design of a unifying language for expressing global constraints, but this language does not necessarily enforce tractability.

The rest of the paper is organized as follows. Section 2 provides the necessary background. Section 3 defines the TFOS problem and proves its tractability. Section 4 describes possible applications of the proposed model. Section 5 provides a scheme for developing a propagation algorithm for TFOS. Section 6 discusses possible extensions of the model. The implications of the TFOS model on the modelling process are discussed in Section 7. A number of concluding remarks are made in Section 8.

2   Background

Given a directed graph $G = (V, E)$ with two specified nodes s and t called *source* and *sink*, a flow in G is a function from the set of arcs $E(G)$ to the set of non-negative integers[1] that satisfies the following conditions: for each vertex v except s and t, the amount of flow entering v equals the amount of the flow leaving v, the amount of the flow entering s as well as the amount of flow leaving t is 0.

---

[1] Generally, a flow does not have to be integral but this restriction is sufficient here.

The maximum flow problem, in its simplest formulation, associates non-negative integer capacities with each edge of G and asks for the maximum flow from s to t such that the flow delivered through each edge does not exceed its capacity. The problem can be solved by picking an initial flow and augmenting it iteratively by finding a path from s to t in the *residual* graph obtained from G by removing some edges and adding "opposites" to other edges (see [1], Sections 6.3 and 6.4 for a detailed description). The time required by each iteration is proportional to the sum of the number of vertices and edges of G. The flow, due to its integrality, is augmented by at least one at each iteration. Hence, the resulting complexity is the complexity of one iteration multiplied by the maximum flow. Polynomial-time algorithms for the maximum flow are well-known [1].

The constraint satisfaction problem (CSP) is defined on a set of variables $VAR = \{var_1, \ldots, var_n\}$ and a set values $VAL = \{val_1, \ldots, val_m\}$. Each variable has a *domain*, which is a subset of VAL. The objective is to assign each variable with exactly one value from its domain subject to certain *constraints*. A constraint specifies a subset S of variables and restricts tuples of values allowed to be assigned to the variables of S. The set S is called the *scope* of the constraint. The scope of a *global* constraint may be of an arbitrary size, even including all the variables. In this paper we consider two types of global constraints: the Global Cardinality Constraint (GCC) [11] and the Among constraint [3]. The former constraint specifies for each value of VAL its minimal and maximal number of occurrences in a solution of the given CSP. The latter constraint specifies for a subset T of VAL the minimal and maximal number of occurrences of values of T in a solution of the given CSP.

The CSP is intractable in general but there have been many tractable classes studied (see, for example, [6]). In particular, a CSP constrained by a single GCC or a single Among constraint is tractable because it can be transformed into a network flow problem [13].

3   The TFOS Model

Let V be a set of vertices. Let F_1 and F_2 be two families of nonempty subsets of V such that any two sets that belong to the same collection are either disjoint or contained one in the other. The intersection between sets from different families may be arbitrary. Each set $Y \in F_1 \cup F_2$ is associated with two non-negative numbers called *minimal and maximal cardinalities* that do not exceed $|Y|$. We refer to the model (V, F_1, F_2) as TFOS, which is an abbreviation of Two Families Of Sets. Let X be a subset of V such that for each element Y of F_1 or F_2, the size of $X \cap Y$ lies between the cardinalities associated with Y^2. We call X a *valid* subset of V. The task of the *TFOS problem* is to find the largest valid subset of V. Consider the following example of application of the proposed model.

Example 1. Consider a scheduling problem with sets J and E of jobs and employees, respectively. Each job is specified by a subset of employees who can perform this job.

---

[2] If $Y \in F_1 \cap F_2$ and, consequently, Y is associated with two distinct pairs of cardinalities, one for F_1, the other for F_2, this condition should be satisfied for both pairs of cardinalities.

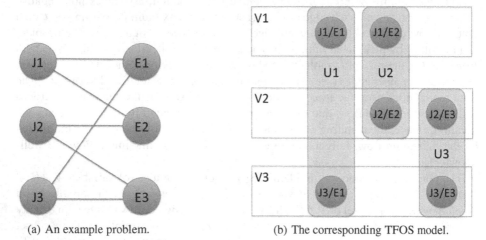

(a) An example problem.   (b) The corresponding TFOS model.

Fig. 1. An illustration of Example 1

Each employee is specified with the minimum and the maximum number of jobs to be performed. The task is to assign each job with exactly one employee so that no employee violates her (or his) restriction of the minimal and maximal allowed number of jobs.

In Figure 1 we have three jobs and three employes. Figure 1(a) shows which job can be performed by which employee: the vertex denoting a job is adjacent to the vertices denoting the employees that can perform this job. The corresponding families of sets are shown in Figure 1(b). In particular, one family of sets denotes sets V_1, V_2, V_3, each of them includes Job/Employee pairs with the same first element. The second family of sets includes sets U_1, U_2, and U_3 which unite the pairs according to the same second element.

More formally, let (V, F_1, F_2) be a TFOS model such that V is the set of all pairs (J_i, E_k) where E_k is an employee who can perform job J_i. Assume that there are n jobs and m employees. Then F_1 contains n subsets of V and the i-th subset contains all pairs with the first element J_i. The cardinalities for each set of F_1 are both 1, expressing the requirement that exactly one employee is to be assigned to a job. The family F_2 contains m subsets of V with the k-th subset containing all pairs having E_k as the second element. The cardinalities of the subset corresponding to employee E_k are the minimal and the maximal number of jobs allowed to E_k. It is not hard to observe that any feasible solution of the specified TFOS problem represents a valid assignment of jobs to employees. Observe that the resulting TFOS model is equivalent to a CSP with a GCC constraint, where F_1 represent domains of variables and F_2 represent cardinality constraints assigned to values.   ▲

Now we prove the tractability of the TFOS problem. Given a TFOS model (V, F_1, F_2), we assume that both F_1 and F_2 cover all vertices of V. If, for example, the set $V \setminus \bigcup F_1$ is not empty, we can add it to F_1 accompanied with cardinalities 0 and $|V \setminus \bigcup F_1|$. Clearly, the resulting TFOS problem is equivalent to the original one.

Let $F_1 = \{S_1, \ldots, S_m\}$, $F_2 = \{T_1, \ldots, T_k\}$. We define the graph $G(F_1, F_2)$ as follows. The vertices of the graph are $s, t, s_1, \ldots, s_m, t_1, \ldots, t_k$, where s_i and t_i correspond to the respective sets, s and t are the source and the sink of the flow. There is an edge (s, s_i), for every S_i which is maximal in F_1 and an edge (t_i, t) for every T_i which is maximal in F_2. There is an edge (s_i, s_j) whenever S_j is a maximal subset of S_i and an edge (t_j, t_i) whenever T_j is a maximal subset of T_i. Finally, let $V(S_i, T_j) \subseteq S_i \cap T_j$ be the set of all u such that S_i and T_j are minimal in their families subject to containing u. There is an edge (s_i, t_j) whenever $V(S_i, T_j)$ is not empty.

Observation 1. *We make the following observations.*

1. *$G(F_1, F_2)$ has exactly one edge entering any s_i and exactly one edge leaving any t_i.*
2. *$G(F_1, F_2)$ has $O(|V|)$ vertices and $O(|V|)$ edges.*

Proof. See Appendix A.

Now we associate with each edge of $G(F_1, F_2)$ its minimal and maximal capacities. The edge entering any s_i or leaving any t_j is associated with the respective minimal and maximal cardinalities of the corresponding set. Finally, the minimal capacity of any edge between s_i and t_j is 0, the maximal capacity is $|V(S_i, T_j)|$.

We will prove that the size of the largest valid subset of V equals the amount of the maximal flow that can be delivered from s to t in $G(F_1, F_2)$. The proof is divided into two lemmas (see Appendix A). In the first one we show that for any valid $X \subseteq V$, there is a flow of size $|X|$. The other lemma shows that for any flow from s to t there is a valid set X whose size equals the amount of the delivered flow. Combining these two lemmas together yields the desired result.

Theorem 1. *Given a TFOS model (V, F_1, F_2), the problem of finding the largest valid subset of V can be solved in $O(|V|^2)$. (In some cases there may be no valid subset at all. In this case, a network flow algorithm reports the absence of feasible flow.)*

Proof. The maximum flow in graph $G(F_1, F_2)$ is at most $|V|$, and all the capacities are integral. Consequently, a traditional iterative approach to solving the maximum flow problem with maximal and minimal capacities (see [1], Section 6.7) solves the problem in $O(|V|)$ iterations. The complexity of each iteration is proportional to the sum of the number of vertices and the number of edges of $G(F_1, F_2)$, which is $O(|V|)$ from Observation 1.   ∎

An illustration of an application of Theorem 1 is presented in Figure 2. In this figure we have a binary CSP: ellipses represent variables, black circles represent the values, and the edges represent the binary conflicts. In this particular example all the conflicts form cliques. So, we can introduce two families of sets $\{V_1, V_2, V_3, V_4\}$ where for each family at least one value has to be selected and the family $\{U_1, \ldots, U_5\}$ of cliques where at most one value can be selected. According to Theorem 1, this CSP can be solved in polynomial time.

Remark. If a TFOS model represents a CSP with n variables and the maximum domain size d then V corresponds to the set of all values of the CSP and has size $O(nd)$; the maximum flow corresponds to a solution of the CSP which has size n. Hence the flow algorithm takes $O(n^2 d)$.

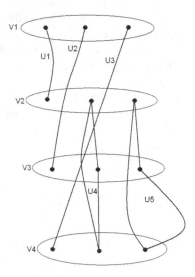

Fig. 2. An illustration of an application of Theorem 1

4   Applications

This section shows that the GCC and the Among constraints, as well as (some of) their combinations, can be represented by the TFOS model. Moreover, it shows how to represent some optimization tasks as a TFOS problem that are hard to express using the classical CSP paradigm.

4.1   GCC and Value-Disjoint among Constraints

Consider the scheduling problem described in Example 1. We specify additional requirements for this problem. Assume that the employees are partitioned according to their professions, based on the minimal and the maximal number of jobs allowed to be performed by persons of each profession. It is not hard to update the TFOS model shown in Example 1 so that it expresses the new requirement. For each profession P, add to F_2 a set that includes all pairs (J_i, E_k) such that E_k has profession P. Set the minimal and maximal cardinalities of that set equal to the minimal and maximal number of jobs, respectively, allowed to fellows of profession P.

It can be shown that this TFOS model is equivalent to a CSP with a combination of GCC and value-disjoint Among constraints [13]. In particular, the value-disjoint Among constraints are represented by the new sets added to F_2. The domains of variables and the cardinality constraints are represented as shown in Example 1.

The resulting TFOS model can be further updated to express new requirements. For example, imagine that the jobs specified in our example scheduling problem span some period of time, say, a week, i.e. the existing constraints restrict the number of jobs in a week. In addition we can restrict the number of jobs performed by each particular person in a day. Let $P_1, \ldots, P_7$ be the partition of jobs according to the day they are to be performed. For each employee E_k, we add to F_2 seven new subsets, the j-th subset contains all elements (J_i, E_k) where J_i is a job of P_j that can be performed by E_k. The cardinalities of j-th subset are the minimal and the maximal number of jobs allowed for E_k on j-th day. Observe that the new elements of F_2 preserve the property of the TFOS model.

4.2   GCC and Variable-Disjoint among Constraints

In the scheduling problem presented in Example 1, assume that only a subset $F \subseteq E$ of employees is constrained by restricting the minimal and maximal number of jobs. Assume, as in the previous subsection, that the jobs of J span a time period of a week and we constrain the number of jobs performed by *all* the employees in a day. That is, we add seven new restrictions to the example problem that specify the minimal and the maximal number of jobs performed by the employees of F in each of the seven days of a week. These new restrictions can be represented as variable-disjoint Among constraints. It can be shown that the resulting problem is equivalent to the combination of GCC and Among constraints if the number of partition classes of J is not necessarily seven, but an arbitrary integer.

The description of the obtained scheduling problem in terms of the TFOS model is not straightforward, because the sets corresponding to the variable-disjoint Among constraints cannot be added to any of the families of sets defined in Example 1 without violating the properties of these families. To obtain the description, observe that there are three possible cases of the example scheduling problem. In the first case, for each job there is an "unconstrained" employee that can perform this job. This case is trivial because each job can be assigned to such an employee without violating any constraint. In the second case, $E = F$, that is, there are no unconstrained employees at all. In this case, partition the sets in F_1 into seven classes according to the day the corresponding jobs are assigned. The Among constraints can be expressed by unions of the sets that belong to the same partition class. The cardinalities for each set are the minimal and the maximal number of jobs allowed on the corresponding day. The sets corresponding to Among constraints can be added to F_1 because they do not violate the required property that any two sets of F_1 are either disjoint or contained one in the other.

In the last case of the example scheduling problem, unconstrained employees can perform only a part of the required jobs but not all. In this case, the TFOS model is constructed in two stages. In the first stage, we take the TFOS model described in Example 1 and replace F_1 by the "projection" of F_1 to F. That is, we replace each set in F_1 by a subset that contains all the elements (J_i, E_k) such that $E_k \in F$. We associate the sets that have lost some of their elements as a result of this replacement with cardinalities 0 and 1, the cardinalities of the other sets both remain 1. In the second stage the Among constraints are introduced by analogy with the previous case. It can be shown that if the resulting TFOS problem does not have a feasible solution, the example scheduling problem does not have a solution either. Otherwise, take any valid set of the obtained TFOS model and assign the jobs "unassigned" by this set to respective unconstrained employees. The resulting assignment is a solution to the problem.

4.3   Compact CSP Modeling

Consider again the scheduling problem and the corresponding TFOS model described in Section 4.1. Recall that the sets of family F_1 represent domains of the corresponding CSP. A relative inflexibility of CSP as a modelling language is that each variable must be assigned with exactly one value from its domain. In this subsection we demonstrate that the TFOS model does not have this disadvantage. In particular we show that by introducing a number of additional subsets into family F_1 of the TFOS model being considered, we can represent a scheduling problem that seems difficult to be expressed in terms of classical CSP.

Recall that the scheduling problem described in Section 4.1 assigns exactly one employee to each job and that the set of all employees is partitioned according to their profession. Assume that we would like to assign each job not with a single employee but with a crew of employees with a specified minimal and maximal number of fellows of each profession participating in the crew. To introduce these new constraints into the TFOS model, we partition each set S of F_1 according to the profession of the employees with which the elements of S correspond. We associate each partition class with the cardinalities equal to the minimal and the maximal number of fellows of the corresponding profession allowed to be assigned to the job corresponding to S. Then we add the resulting new sets to F_1.

5   Towards a Propagation Algorithm for TFOS

In this section we present a Propagation Algorithm (PA) for the TFOS problem. The input of the algorithm consists of a TFOS model (V, F_1, F_2) and an integer k. If there is no valid subset of V of size at least k, the algorithm reports infeasibility. Otherwise, it outputs the subset of V containing the elements that *do not* belong to any valid subset of size at least k. Before presenting the PA itself, let us specify how it can be used to speed up the search.

We assume that an optimization problem is formulated as finding a largest subset of V subject to certain restrictions, a part of which are families F_1 and F_2 with their minimal and maximal cardinalities, and that the problem is solved by a systematic Search Algorithm (SA). In every iteration, the SA possesses additional data: the size m of the largest known subset of V satisfying all the restrictions and a subset V' of V. The SA tries to extend V' to a size of at least $m + 1$. The PA decides whether the extension is possible if the only restrictions considered are those imposed by families F_1 and F_2. If yes, the PA specifies which elements of $V \setminus V'$ cannot participate in such a set. Clearly, if the PA reports infeasibility, the SA must backtrack immediately. If a set of infeasible elements is specified, the SA discards them in its attempt to extend V', thus pruning the branches of the search tree. Note that the PA considers the TFOS model $(V \setminus V', F_1', F_2')$, where the elements of F_1' and F_2' are obtained from F_1 and F_2, respectively, by appropriate restriction of their sets and updating cardinalities. In particular, a set $S \in F_1$ is transformed into a set $S \setminus V'$ of F_1' with the cardinalities $max(l(S) - |S \cap V'|, 0)$ and $u(S) - |S \cap V'|$, where $l(S)$ and $u(S)$ are the cardinalities of S in F_1. The transformation from F_2 to F_2' is analogous. The minimal size of a valid subset "pursued" by the

PA is $m - |V'| + 1$. (If the problem being considered is a CSP, the minimal size of a valid subset always equals to the number of unassigned variables.)

Given a TFOS model (V, F_1, F_2) and an integer k as input, the PA proceeds in two stages. In the first stage, the feasibility of a valid subset of size k is checked. In particular, the PA constructs a graph $G'(F_1, F_2)$ obtained from $G(F_1, F_2)$ by adding a new node s' and an additional edge (s', s) of capacity k. Then the maximum flow from Y s' to t is computed. Clearly a valid subset of size k is feasible only if a flow of size k can be delivered from s' to t. If a valid set of the required size is found out to be feasible, the algorithm goes on to the second stage: computing the subset of infeasible values of V.

Proposition 1. *Let u be a value of V. Let $F_1(u)$ and $F_2(u)$ be the minimal elements of F_1 and F_2 that contain u. Let e be an edge of $G'(F_1, F_2)$ from the node corresponding to $F_1(u)$ to the node corresponding to $F_2(u)$.[3] There is a feasible subset X of V such that $|X| \geq k$ and $u \in X$ if and only if there is flow of size at least k from s' to t and the flow delivered through edge e is nonzero.*

Thus, the set of the infeasible elements of V can be extracted in $O(|V|)$ from the set of *infeasible* edges of $G'(F_1, F_2)$, i.e., the edges that are left "untouched" by any maximum flow from s' to t. These edges can be found by an approach suggested in [11]. According to that approach we consider the residual graph G_R obtained from $G'(F_1, F_2)$ by delivering flow Y. The graph G_R is partitioned into strongly connected components. The infeasible edges are those whose ends do not belong to the same component. Partitioning into strongly connected components for G_R can be done in $O(|V|)$ applying an algorithm by Tarjan (see, for example, [5]). Hence the complexity of the propagation algorithm is determined by the time complexity of the maximum flow computation, which is $O(|V|^2)$ by Theorem 1. Finally, note that the if the TFOS model being considered represents a CSP, the complexity of the PA, which, in terms of CSP, is called *achieving generalized arc-consistency*, is $O(n^2 d)$.

6   Extensions of the TFOS Model

6.1   The Weighted TFOS Problem

Given a TFOS model (V, F_1, F_2) and a weight function w associating each element of V with a weight, the task of the weighted TFOS problem is to find the largest valid subset of V having the smallest weight (the weight of a set is computed as the sum of weights of its elements). By analogy with the unweighted TFOS problem, it can be shown that the weighted TFOS problem generalizes various network flow-based global constraints with costs [12].

Theorem 2. *The weighted TFOS problem is tractable.*

Proof. The weighted TFOS problem can be transformed into the problem of finding the minimum cost flow ([1], Chapter 10) in a graph $G_w(F_1, F_2)$ that can be obtained from

---

[3] There is such an edge because $u \in V(F_1(u), F_2(u))$, hence $V(F_1(u), F_2(u))$ is not empty and thus corresponds to an edge by construction of $G(F_1, F_2)$.

$G(F_1, F_2)$ by introducing the following modifications. The edges entering the nodes corresponding the elements of F_1 or leaving the nodes corresponding to the elements of F_2 are associated with zero costs. Each edge between a node corresponding to a set $A \in F_1$ and a node corresponding to a set $B \in F_2$ is split into $|V(A, B)|$ edges corresponding to elements of $V(A, B)$. The edge corresponding to each element $u \in V(A, B)$ is associated with cardinalities 0 and 1 and with cost $w(u)$. The correctness of the transformation can be proved in a way similar to that employed in Lemmas 1 and 2 (presented in the appendix). ∎

6.2 Three Families of Sets Cause NP-Hardness

If we allow more than two families with the property of families of a TFOS model, the corresponding optimization problem can be shown to be NP-hard.

Theorem 3. *An extension of the TFOS problem that includes three families of sets is NP-hard.*

Proof. The NP-hardness can be shown by the reduction from a version of 3-SAT where each variable occurs at most one as a positive literal and and most twice as a negative literal. This problem is well known to be NP-complete (see, for example, the classical "Computational Complexity" book of Papadimitriou). Let F be a 3-CNF formula over a set of variables $v_1, \ldots v_n$ such that each v_i appears at most once and each $\neg v_i$ appears at most twice. Let Z be a binary CSP with variables corresponding to the clauses of F, the values of the domain of each variable correspond to the literals of the respective clause, two values are incompatible if they correspond to the positive and the negative literal of the same variable. It is not hard to see that Z is soluble if and only if F is satisfiable.

Observe that Z has two types of conflicts. A conflict of the first type can be called an *isolated* conflict. It involves a pair of values (val_1, val_2) which are incompatible but no other value is incompatible with val_1 nor with val_2. A conflict of the second type can be called a *complex* conflict. It involves a triple (val_1, val_2, val_3) such that val_1 is incompatible with both val_2 and val_3 and no other value is incompatible with either of val_1, val_2, val_3.

We introduce the three families of sets structure (U, F_1, F_2, F_3) as follows. The set U includes all the domain values of Z (the values of different domains are considered distinct). F_1 is the family of all domains. The set F_2 is constructed as follows. For each isolated conflict (val_1, val_2), the set $\{val_1, val_2\} \in F_2$. For each complex conflict (val_1, val_2, val_3), the set $\{val_1, val_2)\}$ belongs to F_2. No other sets are contained in F_3. The set F_3 contains only sets $\{val_1, val_3\}$ for each complex conflict (val_1, val_2, val_3). It is not hard to observe that the sets within each family are pairwise disjoint. The associate the upper and lower bounds 1 with the sets of F_1, upper bound 1 and lower bound 0 are associated with the sets of F_2 and F_3.

Observe that (U, F_1, F_2, F_3) has a valid set of size $n = |F_1|$ if and only Z is soluble. Indeed, if Z is soluble then any solution of Z is a valid set of (U, F_1, F_2, F_3) because it has exactly one value within each domain (i.e. each set of F_1) and contains no pairs of conflicting values (i.e has at most one value within each set of F_2 and F_3). Conversely, any valid set has exactly one value in each domain and satisfies all the constraints of Z, hence it is a solution of Z. ∎

6.3 Preserving Tractability: Restrictions on the Third Family

Although the optimization problem based on three families of sets is NP-hard in general, it could be solved efficiently if we restrict the properties of the third family. The following example demonstrates this possibility. We define the three family of sets problem (3FOS) as a four tuple (V, F_1, F_2, F_3), where the first three components are the same as in the TFOS model and F_3 is a family $\{Y_1, \ldots Y_l\}$ of subsets of i such that $Y_i \subset Y_j$ whenever $i > j$. Assume that the minimal cardinalities associated with the elements of F_3 are all zeros. Let $u_1, \ldots u_l$ be the respective maximal cardinalities. We may assume that $u_i < u_j$ whenever $i > j$ as if not, the cardinality constraint imposed by u_i is redundant.

Theorem 4. *The* 3FOS *problem can be solved efficiently using an algorithm for the weighted TFOS problem as a procedure.*

Proof. We associate the elements with weights as follows. All elements of $V \setminus Y_1$ are associated with zero costs. All elements of $Y_1 \setminus Y_2$ are associated with some large positive weight W, say 1000. For $i > 1$, let K be the weight associated with the elements of $Y_{i-1} \setminus Y_i$. Then the elements of $Y_i \setminus Y_{i+1}$ (or Y_i in case $i = l$) are associated with weight $K * u_{i-1}/u_i$. Observe that $X \subseteq V$ violates the cardinality constraints imposed by F_3 if and only if the weight of X is greater then $W * u_1$. This observation suggests the following way of solving the problem.

Solve the weighted TFOS problem (V, F_1, F_2) with the weights assigned as shown below. If there is no feasible valid subset of V then the original problem has no feasible valid subset either. If the resulting largest valid subset X has a weight smaller than or equal to $W * u_1$ then X is a solution of the original problem. Otherwise, we learn that the original problem has no solution of size $|X|$. To introduce this additional constraint, we add to F_1 set V with cardinalities 0 and $|X| - 1$. (If such a set already exists, we adjust its maximum cardinality.) Then we solve the resultant weighted TFOS model again. This process may be repeated a number of iterations. It stops if in some iteration a feasible solution of weight at most $W * u_1$ is found or infeasibility is reported. Otherwise, the maximal allowed cardinality of set V in family F_1 is decreased by 1. Infeasibility is reported if this maximal cardinality has been reduced to 0 with no feasible solution found before. ∎

Although the described example is rather artificial, it it shows the existence of a non-trivial polynomially solvable optimization problem based on more than two families of sets.

7 Modelling Intractable Problems Using the TFOS Model

In this section we demonstrate that the TFOS model can be useful for modelling intractable problems. The main benefit of the TFOS model is that it allows us to model intractable problems in a way that makes constraint propagation more efficient. An intractable problem is usually modelled as a *conjunction* of global constraints [13,14].

Each constraint in the conjunction is propagated separately in a number of iterations until no value can be removed from the domains of the constrained variables. Hence, the number of constraints in the conjunction has a multiplicative factor on the complexity of the propagation algorithm and it is desirable that the number of such constraints be as small as possible. We show now that there are hard problems that can be modelled using a much smaller number of TFOS structures as compared to other types of global constraints.

Consider, for example, the problem obtained at the end of Section 4.1. One can imagine that this problem represents one *shift* of some timetabling problem that can occur in real applications. A timetabling problem usually consists of a number of shifts with intersecting sets of jobs associated with different shifts that make the problem hard [7]. It follows from the description in Section 4.1 that such a hard timetabling problem can be modelled using *one* TFOS structure per shift, while the number of GCC and Among constraints is linear in the number of variables participating in a shift (a linear number of among constraints is needed, for instance, to represent constraints added at the end of Section 4.1). Moreover, the separate propagation of the GCC and Among constraints cannot discard all the values that might be discarded by the propagation of TFOS structures.

8 Conclusion

We have presented an optimization problem that we termed the TFOS problem. We showed that various combinations of network flow-based global constraints can be expressed in terms of the TFOS problem. We also demonstrated that the TFOS model can describe scenarios that seem difficult to express in terms of a classical CSP. We presented global constraints in terms of a scheduling problem rather than an abstract setting, which demonstrated that the TFOS model can be useful for modelling sophisticated resource allocation tasks. We identified some tractable and intractable extensions of the TFOS model. In particular, we showed that the weighted TFOS problem is tractable and that with three families of sets, though intractable in general, it can be made tractable by applying restrictions on the properties of the third family. Finally, we discussed the implications of the TFOS model on the modelling process.

While this paper has raised a route for generalizing collections of flow-based global constraints, much remains to be done. Firstly, it would be interesting to implement our proposed framework and compare its efficiency against more standard approaches to solving collections of flow-based global constraints. Secondly, while we have outlined a propagation scheme for TFOS, a generic filtering algorithm must be developed in order to make the approach more general. A possible direction of further theoretical research is to identify other tractable generalizations of the TFOS problem and to design methods for coping with intractability (such as approximation or parameterized algorithms) for intractable extensions of the TFOS problem.

Acknowledgements

Razgon and O'Sullivan are supported by Science Foundation Ireland (Grant No. 05/IN/I886). Provan is supported by Science Foundation Ireland (Grant No. 04/IN3/I524).

References

1. Ahuja, R., Magnatti, T., Orlin, J.: Network Flows. Prentice-Hall, Englewood Cliffs (1993)
2. Bacchus, F., Walsh, T.: Propagating logical combinations of constraints. In: IJCAI, pp. 35–40 (2005)
3. Beldiceanu, N., Contjean, E.: Introducing global constraints in CHIP. Mathematical and Computer Modelling 12, 97–123 (1994)
4. Bessière, C., Hebrard, E., Hnich, B., Kiziltan, Z., Walsh, T.: The range and roots constraints: Specifying counting and occurrence problems. In: IJCAI, pp. 60–65 (2005)
5. Cormen, T., Leiserson, C., Rivest, R., Stein, C.: Introduction to Algorithms, 2nd edn. The MIT Press and McGraw-Hill Book Company (2001)
6. Dechter, R.: Constraint Processing. Morgan Kaufmann, San Francisco (2003)
7. Meisels, A., Schaerf, A.: Modelling and solving employee timetabling problems. Annals of Mathematics and Artificial Intelligence 39, 41–59 (2003)
8. Pesant, G.: A regular language membership constraint for finite sequences of variables. In: Wallace, M. (ed.) CP 2004. LNCS, vol. 3258, pp. 482–495. Springer, Heidelberg (2004)
9. Quimper, C.-G., Lopez-Ortiz, A., van Beek, P., Golynski, A.: Improved algorithms for the global cardinality constraint. In: Wallace, M. (ed.) CP 2004. LNCS, vol. 3258, pp. 542–556. Springer, Heidelberg (2004)
10. Régin, J.-C.: A filtering algorithm for constraints of difference in CSPs. In: Proceedings of AAAI, pp. 362–367 (1994)
11. Régin, J.-C.: Generalized arc consistency for global cardinality constraint. In: AAAI/IAAI, vol. 1, pp. 209–215 (1996)
12. Régin, J.-C.: Arc consistency for global cardinality constraints with costs. In: Jaffar, J. (ed.) CP 1999. LNCS, vol. 1713, pp. 390–404. Springer, Heidelberg (1999)
13. Régin, J.-C.: Combination of among and cardinality constraints. In: Barták, R., Milano, M. (eds.) CPAIOR 2005. LNCS, vol. 3524, pp. 288–303. Springer, Heidelberg (2005)
14. Régin, J.-C., Puget, J.-F.: A filtering algorithm for global sequencing constraints. In: Smolka, G. (ed.) CP 1997. LNCS, vol. 1330, pp. 32–46. Springer, Heidelberg (1997)

A   Intermediate Proofs from Observation 1 to Theorem 1

Observation 1. *We make the following observations.*

1. $G(F_1, F_2)$ *has exactly one edge entering any* s_i *and exactly one edge leaving any* t_i.
2. $G(F_1, F_2)$ *has* $O(|V|)$ *vertices and* $O(|V|)$ *edges.*

Proof. Each observation is proven separately.

1. If S_i is maximal then (s, s_i) is the only edge that enters s_i. If S_i is not maximal, assume that there are two sets S_j and S_f such that S_i is a maximal subset of both of them. From the structure of F_1, either $S_f \subseteq S_j$ or $S_j \subseteq S_f$. The former case contradicts S_i being the maximal set contained in S_j. In the latter case, there is the same contradiction regarding S_i and S_f. The proof for t_i is symmetric.
2. The statement regarding the number of vertices easily follows from the observation that the number of sets in F_1 as well as in F_2 is at most $2 * |V| - 1$. This observation can be proven by induction on $|V|$. It is immediate if $|V| = 1$. For $|V| > 1$, F_1 may contain (in the worst case) V itself and a partition of V into n, $(n \geq 2)$ subsets of

sizes $y_1, \ldots y_n$. By the induction hypothesis, i-th partition class together with all its subset sum up to at most $2 * y_i - 1$. Summing these numbers together we get $2 * |V| - n$ subsets that together with V itself are at most $2 * |V| - 1$ subsets.

To prove the upper bound on the number of edges, observe that there is at most one edge entering a node corresponding to an element of F_1 and at most one edge leaving a node corresponding to an element of F_2. It follows that we need to check only the number of edges connecting the nodes corresponding to elements of different families. To this point note that each of these edges corresponds to a non-empty subset of V and that the subsets associated with different edges do not intersect. ∎

Lemma 1. *Let X be a valid subset of V. Then there is a flow of size $|X|$ from s to t.*

Proof. We construct the flow as follows. Associate every edge (s_i, t_j) with the flow $|X \cap V(S_i, T_j)|$. Having associated the edges (s_i, t_j) with the appropriate flows, proceed as follows. Whenever there is vertex s_i such that all the edges leaving s_i have already been associated with their flows and the edge entering s_i has not been yet, associate the edge entering s_i with the flow equal to the sum of flows on the edges leaving s_i. Repeat analogously for t_i with the only difference that the edge leaving t_i is associated with the sum of flows on the edges entering t_i. The obtained assignment of flows guarantees that the flow entering s as well as the flow leaving t is zero and that the flow entering each intermediate node equals the flow leaving that node. In the rest of the proof, we show that the constructed flow "respects" all the capacities and has size $|X|$.

Claim. The flow entering any s_i equals $|S_i \cap X|$ and the flow leaving any t_j equals $|T_j \cap X|$.

Proof. We prove the claim regarding $s_1, \ldots, s_m$; the proof regarding $t_1, \ldots, t_k$ is symmetric. Assume that $s_1, \ldots, s_m$ are ordered in such a way that $i < j$ whenever $S_i \subseteq S_j$. The proof is by induction on this sequence. Let $e_1, \ldots, e_l$ be the edges leaving s_1. Due to minimality of S_1 the heads of the edges correspond to sets $T_1' \ldots T_l'$ of F_2. Denote $X \cap V(S_1, T_i')$ by X_i'. Observe that $X_1', \ldots, X_l'$ is a partition of $X \cap S_1$. Indeed, any two X_y' and X_z' are disjoint because any $u \in X_y' \cap X_z'$ implies that one of T_y' and T_z' is contained in the other one contradicting the minimality assumption for the larger set. For any $u \in X \cap S_1$, the set S_1 is the minimal one that contains u just because it is a minimal set in F_1. Let T' be the minimal set in F_2 that contains u. Clearly, there is an edge between the vertices corresponding to S_1 and T', hence there exists a j such that $T' = T_j'$ and u belongs to X_j'. We have proved that $X_1' \ldots X_l'$ are disjoint and cover all the vertices of $X \cap S_1$. Hence, they form a partition of $X \cap S_1$. Consequently, $|S_1 \cap X| = |X_1'| + \ldots + |X_l'|$. Recall that $|X_j'|$ is exactly the flow assigned to e_j and the flow on the edge entering S_1 is the sum of flows on all e_j. The validity of the claim for S_1 follows immediately.

Consider now s_i for $i > 1$. Let $e_1, \ldots, e_l$ be the edges leaving s_i. The head of every e_j is either some s_y or some t_z. In the former case, let $X_j' = S_y \cap X$, in the latter case let $X_j' = V(S_i, T_z) \cap X$. Similar to the case with S_1, we can show that $X_1', \ldots X_l'$ form a partition of $X \cap S_i$. Observe that the flow assigned to every e_j is exactly $|X_j'|$: if the head of e_j is some s_y, it follows from the induction hypothesis taking into account that $y < i$; if the head of e_j is some t_z, the observation follows from the initial assignment of flows. Taking into account that $|S_i \cap X| = |X'_1| + \ldots + |X'_l|$ and that the flow on the edge entering S_i is the sum of flows on the edges leaving S_i, $|S_i \cap X|$ is exactly the flow on the edge entering s_i. $\qquad\square$

Considering that X is a valid set, it respects the minimal and the maximal cardinalities of all the sets in F_1 and F_2. It follows that the flow assigned to the edge entering each s_i is valid and leaving each t_j is valid. For the edges of type (s_i, t_j), the validity follows by definition of the flow on these edges.

It remains to show that the amount of flow delivered from s to t is exactly $|X|$. To this point observe that the amount of flow leaving s is the sum of flows entering the nodes corresponding to the maximal sets of F_1. Let $S'_1, \ldots, S'_l$ be these maximal sets. Clearly, $|S'_i \cap X|$ is exactly the flow entering each vertex s'_i. Taking into account that each element of X belongs to some S'_i, we obtain $X = |S'_1 \cap X| + \ldots |S'_l \cap X|$. $\qquad\blacksquare$

Lemma 2. *Let F be a valid flow from s to t. Then there is a valid set X such that $|X|$ equals the amount of flow that leaves S.*

Proof. For every edge e between vertices s_i and t_j, fix $X(e) \subseteq V(S_i, T_j)$ such that $|X(e)|$ is the amount of flow on edge (s_i, t_j). There is such an $X(e)$ since the maximal capacity on the edge is $|V(S_i, T_j)|$. Let X be the union of all $X(e)$.

Using the inductive argument analogous to the one used in proof of Lemma 1, we can show that the flow entering every s_i and leaving every t_j equals $|S_i \cap X|$ and $|T_j \cap X|$, respectively. Taking into account that the flow is valid, the number of elements of every S_i and T_j contained in X satisfies their minimal and maximal cardinalities. $\qquad\blacksquare$

A Global Filtration for Satisfying Goals in Mutual Exclusion Networks

Pavel Surynek

Charles University
Faculty of Mathematics and Physics
Malostranské náměstí 2/25, 118 00 Praha 1, Czech Republic
surynek@ktiml.mff.cuni.cz

Abstract. We formulate a problem of goal satisfaction in mutex networks in this paper. The proposed problem is motivated by problems that arise in concurrent planning. For more efficient solving of goal satisfaction problem we design a novel global filtration technique. The filtration technique is based on exploiting graphical structures of the problem - clique decomposition of the problem is used. We evaluated the proposed filtration technique on a set of random goal satisfaction problems as well as a part of GraphPlan based planning algorithm. In both cases we obtained significant improvements in comparison with existing techniques.

Keywords: global filtration, mutual exclusion network, search.

1 Introduction

We propose a new global filtration method for satisfying goals in *mutual exclusion networks* in this paper. The mutual exclusion network is an undirected graph where a finite set of symbols is assigned to each vertex. The interpretation of edges is that a pair of vertices connected by an edge cannot be selected together. In other words, edges in the graph represent mutual exclusions of vertices (or conflicts between vertices). Having a goal, which is a finite set of symbols, the task is to select a stable set of vertices in this graph such that the union of their symbols covers the given goal.

This problem may seem artificial at first sight but in fact it is a slight reformulation of problems that appear in *concurrent planning* for artificial intelligence [1] with planning graphs [2] and in *Boolean formula satisfaction* [3]. In addition to these applications the defined problem may be generic enough to worth studying for itself (the more detailed motivation is given in section 2).

Existing techniques that can be used to solve the problem include variety of backtracking based search methods enhanced with consistency techniques which is a typical approach used in *constraint programming* practice [4] and which we are following too. Our experiments expectably showed that local consistency techniques (namely arc-consistency) can bring significant improvements in terms of overall solving time compared to plain backtracking. Nevertheless, this result invoked a question what an improvement can be obtained by using a certain type of global consistency? We are trying to answer this question in this paper.

F. Fages, F. Rossi, and S. Soliman (Eds.): CSCLP 2007, LNAI 5129, pp. 142–157, 2008.

As a first step we investigated the possible usage of existing global constraints [4, 12] for modeling the problem. We considered several existing global constraints based on network flows (such as *allDifferent* and similar constraints; our idea was to simulate the problem as a network flow and then model it using these constraints). But it turned out that existing global constraints are not suitable for modeling the problem (the reason we identified as a main obstacle is the quite complicated relation between the impact of selection of a vertex on the rest of the mutual exclusion network and the goal we are trying to satisfy). The option was to develop our own specialized global filtering method for the problem which we eventually did.

Our filtering method exploits the structural information encoded in the problem. Valuable structural information in the mutual exclusion network is a *complete subgraph* (clique). More precisely we can extract this structural information if we have a clique decomposition of the network.

If we know that several vertices in the graph form a clique we also know that at most one of them can be selected into the solution. This simple property allows us to do further relaxed reasoning. The clique of vertices can be treated as single entity with a limited contribution to the solution (since only one vertex can be selected which typically means that not a lot symbols in the goal can be covered by the clique). Then we can check relaxed condition on selection of a vertex into the solution. A vertex can be selected into the solution if the maximum number of symbols that can be obtained from the remaining cliques plus the number of symbols of the vertex is not lower than the number of symbols in the goal. This condition is necessary but not sufficient, however if it does not hold we can filter out the vertex from further consideration.

This paper is organized as follows. First we give more details about our motivation to deal with the problem. In the next two sections we introduce some formalism through which we will express the problems and we discuss some complexity issues. The fourth section is devoted to the description of our new global filtration. In the main part, we evaluate our approach using a set of benchmarks. Finally, we put our work into relation with existing works on the similar topic and we sketch out some ideas for future development.

2   Motivation by AI Problems

We would like to give a motivation for studying goal satisfaction problems in mutual exclusion networks in this section. Generally, we have two sources of motivation from artificial intelligence - the first is *concurrent planning* and the second is *Boolean formula satisfaction*.

The main motivation for our work was *concurrent planning* known from artificial intelligence [6,7,9]. To provide better insight into our motivation let us introduce concurrent planning briefly. Although the formalism and theory around concurrent planning is quite complex the basic idea is simple. Let us have a certain planning world (as an example we can consider a planning world shown in the upper part of figure 1). The task we want to fulfill is to transform the given planning world by executing actions from a set of allowed actions into a state that satisfies certain goal condition (as an example of the goal condition for a planning world we can take the planning world shown in the lower part of figure 1). An action in this context is an

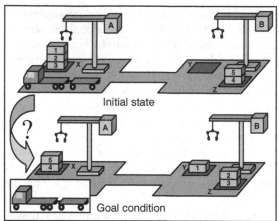

Concurrent plan for reaching the goal
{take(craneA, box1, pileX); take(craneB, box5, pileZ)}
{load(truck, craneA, box1); put(craneB, box5, pileY)}
{move(truck, siteA, siteB); take(craneB, box4, pileZ);
take(craneA, box2, pileX)}
{load(truck, craneB, box4); put(craneA, box2, pileX)}
{take(craneB, box5, pileY); take(craneA, box2, pileX)}
{put(craneB, box5, pileZ)}
{put(craneA, box2, pileX); unload(truck, craneB, box1)}
{move(truck, siteB, siteA); put(craneB, box1, pileY);
take(craneA, box2, pileX)}
{load(truck, craneA, box2); take(craneB, box1, pileY)}
{move(truck, siteA, siteB); put(craneB, box1, pileY);
take(craneA, box3, pileX)}
{put(craneA, box3, pileX); unload(truck, craneB, box2)}
{move(truck, siteB, siteA); put(craneB, box2, pileY);
take(craneA, box3, pileX)}
{load(truck, craneA, box3); take(craneB, box5, pileZ)}
{put(craneB, box5, pileY); unload(truck, craneA, box4)}
{move(truck, siteA, siteB); take(craneB, box5, pileY)}
{load(truck, craneB, box5); put(craneA, box4, pileX)}
{take(craneA, box4, pileX); unload(truck, craneB, box3)}
{move(truck, siteB, siteA); put(craneA, box4, pileX);
put(craneB, box3, pileZ)}
{take(craneB, box2, pileY); unload(truck, craneA, box5)}
{put(craneA, box5, pileX); put(craneB, box2, pileZ)}

Fig. 1. An example of planning problem. The task is to transform the initial state of a given planning world into a planning world satisfying the goal condition (in the goal condition we do not care where the truck is located). A concurrent solution plan is in the right part of the figure.

elementary operation that locally changes the planning world (such an elementary operation in figure 1 is for example *take box 1 by crane A*).

In the basic variant of planning problems we are searching for a sequence of actions that, when executed one by one starting in the given planning world, results into the planning world that satisfies the goal. The concurrent planning itself represents a generalization of this basic variant. Particularly, we allow more than one action to be executed in a single step in concurrent planning. This generalization is motivated by the fact that certain actions do not interfere with each other and they can be executed simultaneously without influencing each other (such non-interfering actions in figure 1 in the upper part are for example *take box 1 by crane A* and *take box 5 by crane B*; the pair of actions *load box 1 by crane A on truck* and *move truck from left location to right location* do interfere). Thus, the task in concurrent planning is to find not just a sequence of actions but a sequence of sets of non-interfering actions that when executed starting in the given planning world results into a goal satisfying state. The execution of a sequence of sets of actions means that we are executing sets of actions one by one where actions from each set are executed simultaneously. This is allowed by the fact that actions in each set of the sequence do not interfere.

But what is the relation between the concurrent planning and the mutual exclusion network mentioned in the introduction? The frequently asked question which arise during solving process of algorithms for concurrent planning is "What are the sets (or is there any) of non-interfering actions that satisfies certain goal?". To be more concrete, this question often arises during the solving process with the usage of the framework of so called *planning graphs* [2]. This is the case of the pioneering *GraphPlan* algorithm as well as of its modern derived variants [8,9,10]. This question can be directly modeled as a mutual exclusion network (actually we borrowed the term mutual exclusion from the planning graph terminology), where the vertices of the network are represented by actions (a set of terms that forms the effect of the action is assigned to the corresponding vertex as a set of symbols) and edges of the network are represented by pairs of actions that interfere with each other.

The minor motivation to study the concept of mutual exclusion networks is *Boolean formula satisfiability*. We found that Boolean formula satisfaction problems (*SAT*) [3, 19] can be modeled as mutual exclusion networks. This issue is studied in more details in [17, 18], therefore we mention it as a motivation only. However, let us note that a SAT problem consists in finding of a valuation of Boolean variables that satisfies a given formula in *conjunctive normal form* (*CNF* - conjunction of clauses where clause is a disjunction of literals).

Intuitively, it is possible to observe that such modeling can be done by declaring literals to be vertices of the network where each vertex (literal) has assigned a set of clauses in which it appear as its set of symbols. The goal would be the set of all the clauses of the given formula and edges in the network would connect vertices (literals) which are conflicting (in the most trivial case, literals x and $\neg x$, where x is a variable, are conflicting).

These two areas of application of our concept of global filtering are especially suitable since they often contain properties of objects that behave like functions. That is, a single value can be assigned to the property of an object or of a group of objects at the moment (for example imagine a robot at coordinates $[3,2]$, the robot can move to coordinates in its neighborhood, so the possible actions are: $moveTo([2,2])$, $moveTo([2,3])$, …; the robot can choose only one of these actions at the moment; executing more than one action at once is physically implausible). Such functional property typically induces a complete sub-graph in the mutual exclusion network.

3   Mutual Exclusion Network and Related Problem

We define mutual exclusion network and problems associated with it formally in this section.

The following definitions formalize mutual exclusion network (shortly *mutex network*) and the associated *problem of satisfying goals* in the mutex network. We assume a finite universe of symbols S for the following definitions.

Definition 1 (Mutual exclusion network). Mutual exclusion network is an undirected graph $N = (V, M)$, where a finite set of symbols $\emptyset \neq S(v) \subseteq S$ is assigned to each vertex $v \in V$. □

Definition 2 (Goal satisfaction in mutex network). Given a goal $G \subseteq S$ and a mutex network $N = (V, M)$ the problem of satisfying goal G in the mutex network N is the task of finding a stable set of vertices $U \subseteq V$ such that $G \subseteq \bigcup_{u \in U} S(u)$. □

An example of mutual exclusion network and a problem of goal satisfaction in this network are shown in figure 2.

The problem of goal satisfaction in mutex network is computationally difficult. To show this claim we can use a polynomial time reduction of the Boolean formula satisfaction problem to the problem of goal satisfaction in mutex network. Then it remains only little to conclude that the problem of goal satisfaction in mutex network is *NP*-complete.

Theorem 1 (Complexity of goal satisfaction). *The problem of goal satisfaction in mutex network is NP-complete.* ∎

Sketch of proof. If we are given a set of vertices we are able to decide whether it is a solution of the problem or not in polynomial time. Hence, the problem is in *NP* class. *NP*-hardness can be proved by using polynomial time reduction of Boolean formula satisfaction problem (*SAT*) to the problem of goal satisfaction in mutex network. Consider a Boolean formula B over a set of Boolean variables. It is possible to assume that the formula B is in the form of conjunction of disjunctions, that is $B = \bigwedge_{i=1}^{n} \bigvee_{j=1}^{m_i} x_j^i$, where x_j^i is a variable or a negation of a variable (literal). For each clause $\bigvee_{j=1}^{m_i} x_j^i$ where $i = 1, 2, \ldots, n$ we introduce a symbol i into the constructed goal G. We introduce vertices v and $\neg v$ into the network for every variable v from the set of variables. A set of symbols $S(x) = \{i \mid x \in \bigcup_{j=1}^{m_i} \{x_j^i\}\}$ is assigned to each vertex x of the network (set of symbols for a vertex corresponds to the set of clauses in which the literal corresponding to the vertex occurs). Finally we add an edge $\{x_j^i, x_l^k\}$ into the mutex network if $x_j^i = v$ and $x_l^k = \neg v$ or $x_j^i = \neg v$ and $x_l^k = v$ for some Boolean variable v (x_j^i and x_l^k are positive and negative literals of the same variable). Now it is sufficient to observe that we can obtain a solution of the original Boolean formula satisfaction problem from the solution of the constructed problem of goal satisfaction in polynomial time. ∎

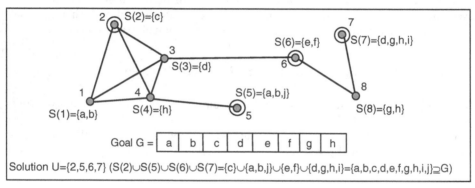

Fig. 2. An instance of the problem of satisfying goal in a mutual exclusion network. The solution is depicted by circles around vertices.

There is little hope to solve the problem of goal satisfaction in mutex network effectively (in polynomial time) in the light of this result. It seems that search is the only option to solve the problem. However, the search may be more or less informed. The more informed search leads to the lower number to steps required for obtaining a solution (the number of steps of the search is usually in tight relation with the overall solving time). One of the most successful techniques how the search can be made more informed is the usage of so called *filtration techniques* (or consistency techniques) which are used intensively in constraint programming [4].

The filtration technique is a specialized algorithm that enforces certain necessary condition for existence of the solution in the problem. Since the basic requirement on the filtration technique is its high speed and low space requirements the necessary conditions that are used in practice represent relatively big relaxations of the original problem. Enforcing of the consistency is done in most cases by ruling out the values from the variables over which the search makes decisions. Such removal of values reduces the size of the search space. The amount of search space reduction is determined by the strength of the filtration technique (that is by the strength of the enforced necessary condition). On the other hand the stronger filtration technique is often redeemed by its higher time complexity. Therefore a balanced trade-off between strength of filtration technique and resource requirements must be found.

A well known example of filtration technique is *arc-consistency* [11]. This is the representative of the technique that enforces certain type of local necessary condition. Locality of filtering technique means that a small number of decision variables is considered at once (in the case of arc-consistency only two variables are considered at once). Another advantage of local filtering techniques is that they are usually highly generic which allows using them in variety of problems with no or little adaptation.

The stronger filtration can be achieved by so called *global filtering techniques*. These methods take into account more than two decision variables at once (in the extreme case all the decision variables in the problem). The large portion of the problem considered at once inherently implies stronger necessary conditions that can be enforced. However, the drawback of global filtering techniques is that they are associated with particular sub-problems (for example the problem where we have several variables with finite domains of values and we require pair-wise different values to be assigned to these variables respecting variable's domains - *allDifferent* filtering technique [12]) which precludes their usage when we cannot recognize the right sub-problem in the problem of our choice.

4 Global Filtration for Goal Satisfaction in Mutex Network

We have the formal definition of the problem we are about to solve it at this point. The solving approach we develop in this section is a new global filtration technique. The technique will be designed specially for problems of goal satisfaction in mutex network (with regard on applications in concurrent planning).

We visually observed that mutex networks obtained from problems arising in concurrent planning embody high density of edges grouped in relatively small number of clusters (this observation was done using our visualization utility on the series of concurrent planning problems). Let us note that our method works with sparse mutex networks as well. The high density of edges and their structural distribution is caused by various factors. Nevertheless, we regard the functional character of properties of objects encoded in the network as the most important one (in planning generally, the functional character of object's properties is typical). Values that form the domain of such property induces complete sub-graphs (clique) in the mutex network. The described structural characterization of mutex networks we can meet in concurrent planning can be exploited for designing of a filtration technique.

If we know a clique decomposition of the mutex network we can reason about the impact of the vertex selection on possibility of goal satisfaction. To be more concrete, we know that at most one vertex from each clique of the decomposition can be selected to contribute to the satisfaction of the goal. Hence, for each clique of the decomposition we can calculate the maximum number of symbols of the goal which can be covered by the vertices of the clique. When we select a vertex into the solution the necessary condition on the solution existence is that the number of symbols covered by the remaining cliques of the decomposition together with symbols associated with selected vertex must not be lower than the number of symbols in the goal.

The second part of the idea of our filtration technique is that if we restrict ourselves on the proper subset of the goal the set of vertices ruled out by the above counting arguments can be different. Therefore it is possible to perform filtration by the technique with respect to multiple sub-goals of the goal to achieve the maximum pruning power. We call these sub-goals of the goal *projection goals* and according to this designation we call the whole filtration technique *projection consistency*.

4.1 Partitioning the Mutex Network into Cliques

Projection consistency assumes that a partition into cliques of a mutex network is known. Thus we need to perform a preprocessing step in which a partition into cliques of the mutex network is constructed. Let $N = (V, M)$ be a mutex network. The task is to find a partition of the set of vertices $V = C_1 \cup C_2 \cup ... \cup C_n$ such that $C_i \cap C_j = \emptyset$ for every $i, j \in \{1, 2, ..., n\} \& i \neq j$ and C_i is a clique with respect to M for $i = \{1, 2, ..., n\}$. Cliques of the partitioning do not cover all the edges in the network in general case. For $m = M - (C_1^2 \cup C_2^2 \cup ... \cup C_n^2)$, $m \neq \emptyset$ holds in general (where $C^2 = \{\{a, b\} \mid a, b \in C \& a \neq b\}$). Our requirement is to minimize n and $|m|$ somehow. Unfortunately this problem is too hard for reasonable objective functions of n and $|m|$ to be solved within the preprocessing step (for instance it is *NP*-complete for minimizing just n [5]).

As an exponential amount of time spent in preprocessing step is unacceptable it is necessary to abandon the requirement on optimality of partition into cliques. It is sufficient to find some partition into cliques to be able to introduce projection consistency. Our experiments showed that a simple greedy algorithm provides satisfactory results. Its complexity is polynomial in size of the input graph which is acceptable for the preprocessing step. The greedy algorithm we are using repeatedly finds the largest greedy clique; the clique is extracted from the network in each step; the algorithm continues until the network is non-empty. For detailed description of this process see [14]. We also made some experiments with partition into cliques of higher qualities than that produced by the greedy algorithm. However, we did not observe any subsequent improvement of the filtering strength of projection consistency.

4.2 Formal Definition of Projection Consistency

For the following formal description of projection consistency we assume that a partition into cliques $V = C_1 \cup C_2 \cup ... \cup C_n$ of the mutex network $N = (V, M)$ was constructed. Projection consistency is defined over the above clique decomposition for a projection goal $\emptyset \neq P \subseteq G$. The projection goal P enters the definitions as a parameter. Projection goals are used for restricting the consistency on a certain part of the goal satisfaction problem (on certain part of the goal) which may eventually strengthen the necessary condition we are about to check.

The fact that at most one vertex from a clique can be selected into the solution allows us to introduce the following definition.

Definition 3 (Clique contribution). A *contribution of a clique* $C \in \{C_1, C_2, \ldots, C_n\}$ to the projection goal $\emptyset \neq P \subseteq G$ is defined as $\max(|S(v) \cap P| \mid v \in C)$ and it is denoted as $c(C, P)$. □

The concept of clique contribution is helpful when we are trying to decide whether it is possible to satisfy the projection goal by selecting the vertices from the partition into cliques. If for instance $\sum_{i=1}^{n} c(C_i, P) < |P|$ holds then the projection goal P cannot be satisfied. Nevertheless, the projection consistency can handle a more general case as it is described in the following definitions.

Definition 4 (Projection consistency: supported vertex). A vertex $v \in C_i$ for $i \in \{1, 2, \ldots, n\}$ is *supported* with respect to a given clique decomposition and the projection goal P if $\sum_{j=1, j \neq i}^{n} c(C_j, P) \geq |P - S(v)|$ holds. □

Definition 5 (Projection consistency: consistent problem). An instance of the problem of satisfaction of a goal G in a mutex network $N = (V, M)$ is consistent with respect to the given clique decomposition and the projection goal $\emptyset \neq P \subseteq G$ if every vertex $v \in C_i$ for $i = 1, 2, \ldots, n$ is supported with respect to the given clique decomposition and the given projection goal. □

It is easy to observe that projection consistency is a necessary but not sufficient condition on existence of the solution. This claim is formally proved in [15].

Algorithm 1: Projection consistency propagation algorithm

---

function *propagateProjectionConsistency* $(\{C_1, C_2, \ldots, C_n\}, P)$: **set**
1: $\gamma \leftarrow 0$
2: **for** $i = 1, 2, \ldots, n$ **do**
3: │ $c_i \leftarrow$ *calculateCliqueContribution* (C_i, P)
4: │ $\gamma \leftarrow \gamma + c_i$
5: **for** $i = 1, 2, \ldots, n$ **do**
6: │ **for each** $v \in C_i$ **do**
7: │ │ **if** $\gamma + |S(v) \cap P| < |P - S(v)| + c_i$ **then** $C_i \leftarrow C_i - \{v\}$
8: **return** $\{C_1, C_2, \ldots, C_n\}$

function *calculateCliqueContribution* (C, P) : **integer**
9: $c \leftarrow 0$
10: **for each** $v \in C$ **do**
11: │ $c \leftarrow \max(c, |S(v) \cap P|)$
12: **return** c

A propagation algorithm for projection consistency is shown here as algorithm 1. Clique decomposition and projection goal are parameters of the algorithm. The algorithm runs in $O(|V| |P|)$ steps which is polynomial in size of the input [15].

To ensure maximum vertex filtration effect we can enforce the consistency with respect to multiple projection goals. However, it is not possible to use all the projection goals since they are too many. Our experiments showed that projection goals P_i

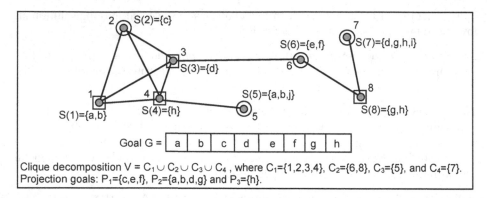

Fig. 3. Example of projection consistency enforcing. The goal satisfaction problem is same as in figure 2. Unsupported vertices are surrounded by squares. For example vertex 3 is unsupported for the projection goal $P_1=\{c,e,f\}$ since vertex 3 contributes by 0, C_2 contributes by 2, C_3 contributes by 0, and C_4 contributes by 0 which is together less than the size of P_1.

where $P_i = \{s \mid s \in G \,\& \left| \{v \mid s \in S(v) \cap G\} \right| = i\}$ provide satisfactory filtration effect (precisely, it is the best selection rule we found by experimentation). The number of projection goals of this form is linear is size of the goal G.

An example of projection consistency enforcing in the goal satisfaction problem from figure 2 is shown in figure 3.

5   Experimental Evaluation

This section is devoted to experimental evaluation of the projection consistency. Our experimental evaluation is concentrated on two aspects of the proposed global consistency. Firstly, we would like to evaluate the consistency itself by using a set of randomly generated goal satisfaction problems. Secondly, we would like to evaluate the benefit of the new consistency when it is applied in concurrent planning. We carried out this evaluation by integrating the consistency into the GraphPlan based algorithm for generating concurrent solutions of planning problems.

5.1   Random Goal Satisfaction Problems

When we visually observed how do the problems arising in concurrent planning look like the distribution of structures was clearly evident. The mutex network associated with the problems typically consists of small number of relatively large cliques accompanied with small number of edges not belonging to any clique. The example of such mutex network is shown in figure 4.

The most variable part of the problem as it was evidenced by our observation is the number of edges not belonging to any clique. Therefore we decided to have this parameter as the main variable parameter in our set of randomly generated problems.

As a competitive technique we chose arc-consistency since it is similar to our new technique in several aspects. First, arc-consistency is easy to implement. This is also

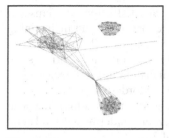

Fig. 4. Mutex network arising as a sub-problem during concurrent solution construction of a planning problem

true for projection consistency. Second, both filtration techniques remove values from the decision variables (not tuples of values etc.).

We integrated both techniques into a backtracking based algorithm for solving the goal satisfaction problem in mutex network. The algorithm performs filtration after each decision - so we are maintaining arc-consistency [11] or projection consistency respectively in our solving approach.

We evaluated our global filtration technique in comparison with arc-consistency on a set of random problems of goal satisfaction of the following setup

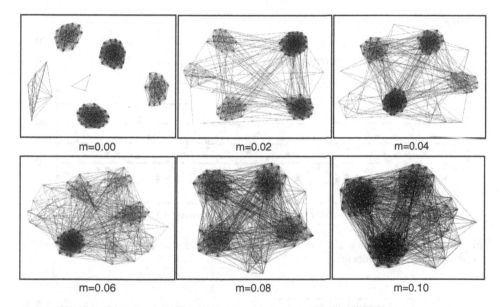

Fig. 5. Random mutex networks with 120 vertices and with fixed structured part (several complete sub-graphs) and with increasing portion of randomly added edges. The parameter m is the probability of presence of an edge between a pair of vertices. Mutex networks shown in this figure were used for experimental evaluation.

motivated by the visual observation of problems arising in concurrent planning. In a mutex network consisting of 120 vertices we constructed several complete sub-graphs using uniform distribution with the mean value of 20.0. The size of the goal was 60 and each vertex has assigned a random set of symbols from the goal of the size generated by the normal distribution with the mean value of 8.0 and the standard deviation of 6.0. Finally we added random edges into the mutex network. More precisely, we add each possible edge into the mutex network with the probability of m where m was a variable parameter which ranged from 0.0 to 0.1. The illustration of randomly generated mutex networks used in our evaluation is shown in figure 5.

For each value of the parameter m we generated 10 goal satisfaction problems and we solved them using backtracking with maintaining arc-consistency and maintaining projection consistency respectively. Along the solving process we collected data such as number of backtracks, runtime etc. The variable and value ordering heuristics are the following. A variable with the smallest domain (smallest clique) is selected preferably. Values (vertices) within the variable's domain are not ordered.

The tested algorithms were implemented in C++ and were run on a machine with AMD Opteron 242 processor (1.6 GHz) and 1 GB of memory under Mandriva Linux 10.2. The code was compiled by gcc compiler version 3.4.3.

For each value of parameter m we calculated average runtime of both techniques, runtime of the easiest problem (the problem with the fewest number of backtracks) and the runtime of the hardest problem (the problem with the highest number of backtracks). The results we obtained are shown in figures 6, 7, and 8.

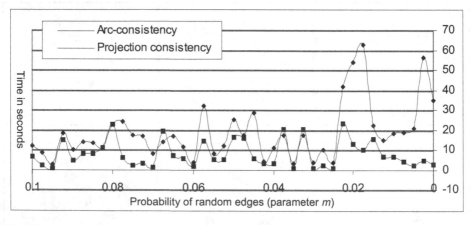

Fig. 6. Runtime for random goal satisfaction problems (average of 10 problems for each value of random edge probability m)

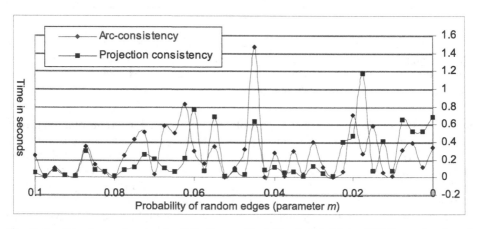

Fig. 7. Runtime for random goal satisfaction problems (easiest problem of 10 for each value of random edge probability m)

The results show that backtracking with maintaining projection consistency is generally faster than backtracking with maintaining arc-consistency on a set of tested problems. In some cases, version with maintaining projection consistency is several times faster (figure 6). The version with maintaining projection consistency achieves better improvement compared to the version with maintaining arc-consistency on harder problems (figure 8). On the other hand, on easy problems projection consistency provides little or no advantage (figure 7). We may also observe that harder problems tends to occur more for lower values of m. On these problems projection consistency represents clearly the better option.

The improvement ratio of solving algorithm using projection consistency with respect to the version with maintaining arc-consistency is shown in figure 9. On the tested problems we reached the improvement up to the order of magnitude.

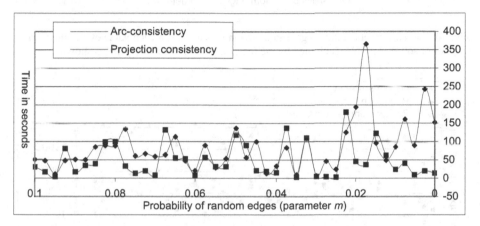

Fig. 8. Runtime for random goal satisfaction problems (hardest problem of 10 for each value of random edge probability m)

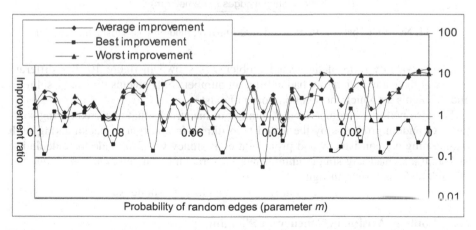

Fig. 9. Improvement ratio of backtracking with maintaining projection consistency with respect to backtracking with maintaining arc-consistency on random goal satisfaction problems

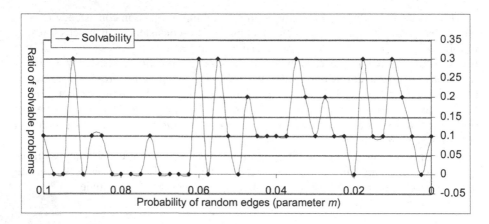

Fig. 10. Ratio of solvable random goal satisfaction problems. For each value of parameter m the number of solvable problems divided by the total number of problems (=10) is shown.

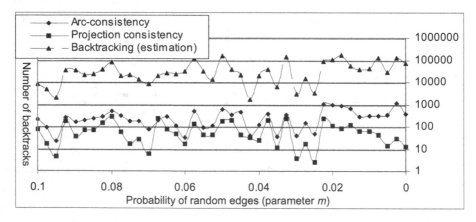

Fig. 11. Number of backtracks of tested algorithms on random goal satisfaction problems

The solvability ratio of the tested problems is shown in figure 10. For different values of the parameter m we had a different numbers of problems that had a solution and problems for which the solution does not exist.

We also performed comparison of number of backtracks that occurred during solving the random problems by the tested algorithms. In addition to backtracking with maintaining arc-consistency and projection consistency we also made the calculation of backtracks made by simple uninformed backtracking. The comparison of number of backtracks is shown in figure 11. According to figure 6 and figure 11 we can observe that the runtime and the number of backtracks correspond well.

5.2 Problems Arising in Concurrent Planning

We also evaluated the proposed projection consistency in solving problems that arise in concurrent planning (that is in the area for which the filtering technique was designed). We used GraphPlan planning algorithm [2] for this evaluation. This algorithm often solves a sub-problem that can be reformulated to a goal satisfaction problem in mutex network.

In our evaluation we used maintaining arc-consistency and projection consistency respectively to improve solving process of this sub-problem within the planning algorithm. We used a set of planning problems of three domains - dock worker robots domain, towers of Hanoi domain, and refueling planes domain. The tested problems were of various difficulties. The length of solution plans ranged from 4 to 44 actions. The comparison of runtime of standard GraphPlan and versions enhanced with maintaining arc-consistency and projection consistency is shown in figure 12. All the planning problems used in this evaluation are available at the web site: http://ktiml.mff.cuni.cz/~surynek/research/rac2007/ (we use our own format of planning problems since we use non-standard representation with explicit state variables).

The improvement obtained by using projection consistency is up to 1000% with respect to both the standard GraphPlan as well as with respect to the version enhanced by arc-consistency. Additionally, we found that goal satisfaction problems arising in these planning problems are very similar to random goal satisfaction problems with the parameter ranging from 0.07 to 0.02 where the improvement obtained by projection consistency is promising.

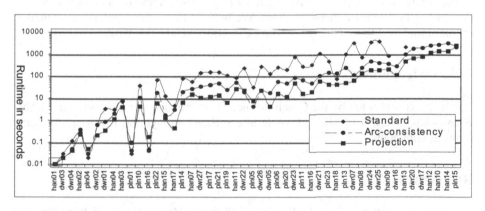

Fig. 12. Runtime comparison of GraphPlan based planning algorithm with versions of this algorithm enhanced by maintaining arc-consistency and projection consistency for solving goal satisfaction problems on a set of planning problems of various difficulties

6 Note on Additional Related Works and Conclusion

Originally, we proposed projection consistency in [15]. This paper is dedicated to theoretical properties of the technique (theoretical comparison with arc-consistency is given in the paper). A study of using similar technique to projection consistency in solving difficult Boolean formula satisfaction problems is given in [17]. The application of arc-consistency in planning using planning graphs is proposed in [16]. We also investigated the possible strengthening of the projection consistency by replacing the expression $\sum_{j=1, j \neq i}^{n} c(C_j, P) \geq |P - S(v)|$ in the definition 4 by $\sum_{j=1, j \neq i}^{n} c(C_j, P - S(v))$

$\geq |P - S(v)|$. Details are given in [13]; but let us note that this change complicates effect of vertex removal too much in a general network (the so called *monotonicity* [15] does not hold).

The ideas of using constraint programming techniques in concurrent planning are presented in [6, 7]. However, only local propagation techniques are studied there (contrary to our approach which is global). Concurrent planning itself is studied in [20]. In this work we were primarily inspired by global constraints such as that presented in [12].

We proposed a novel global filtration technique for a problem of goal satisfaction in mutual exclusion networks. The problem for which the filtration was designed was inspired by concurrent planning. However, the technique is more general, currently we know that it is also effectively applicable in solving SAT problems.

We evaluated our technique in comparison with arc-consistency on a set of randomly generated problems. For this evaluation we used our own implementation in C++. The improvement gained by using projection consistency is up to the 10 times shorter runtime; a similar improvement was achieved in number of backtracks. Finally, we integrated our technique into the GraphPlan planning algorithm to evaluate it in some area of application. Again we obtained significant improvements compared to the standard version.

For future work we plan to investigate more precise computation of supported vertices (definition 4) using network flows. We believe that a more precise computation of this would lead to a stronger necessary condition we are checking.

Acknowledgements

This work is supported by the Czech Science Foundation under the contracts number 201/07/0205 and 201/05/H014. I would like to thank anonymous reviewers for many rigorous comments.

References

1. Allen, J., Hendler, J., Tate, A.: Readings in Planning. Morgan Kaufmann Publishers, San Francisco (1990)
2. Blum, A.L., Furst, M.L.: Fast Planning through planning graph analysis. Artificial Intelligence 90, 281–300 (1997)
3. Cook, S.A.: The Complexity of Theorem Proving Procedures. In: Proceedings of the 3rd Annual ACM Symposium on Theory of Computing, pp. 151–158. ACM Press, USA (1971)
4. Dechter, R.: Constraint Processing. Morgan Kaufmann Publishers, San Francisco (2003)
5. Golumbic, M.C.: Algorithmic Graph Theory and Perfect Graphs. Academic Press, London (1980)
6. Kambhampati, S.: Planning Graph as a (Dynamic) CSP: Exploiting EBL, DDB and other CSP Search Techniques in Graphplan. JAIR 12, 1–34 (2000)
7. Kambhampati, S., Parker, E., Lambrecht, E.: Understanding and Extending GraphPlan. In: Steel, S. (ed.) ECP 1997. LNCS, vol. 1348, pp. 260–272. Springer, Heidelberg (1997)

8. Koehler, J.: Homepage of IPP. Research web page University of Freiburg, Germany (April 2007), http://www.informatik.uni-freiburg.de/~koehler/ipp.html
9. Little, I., Thiébaux, S.: Concurrent Probabilistic Planning in the Graphplan Framework. In: Proceedings of the 16th International Conference on Automated Planning and Scheduling, Cumbria, UK, pp. 263–272. AAAI Press, Menlo Park (2006)
10. Long, D., Fox, M.: Efficient Implementation of the Plan Graph in STAN. JAIR 10, 87–115 (1999)
11. Mackworth, A.K.: Consistency in Networks of Relations. Artificial Intelligence 8, 99–118 (1977)
12. Régin, J.C.: A Filtering Algorithm for Constraints of Difference. In: Proceedings of the 12th National Conference on Artificial Intelligence, pp. 362–367. AAAI Press, Menlo Park (1994)
13. Surynek, P.: Tractable Class of a Problem of Goal Satisfaction in Mutual Exclusion Network. In: Proceedings of the 21st International FLAIRS Conference, Miami, FL, USA. AAAI Press, Menlo Park (to appear, 2008)
14. Surynek, P.: Projection Global Consistency: An Application in AI Planning. Technical Report, ITI Series, 2007-333. Charles University, Prague (2007), http://iti.mff.cuni.cz/series
15. Surynek, P.: Projection Global Consistency: An Application in AI Planning. In: Proceedings of CSCLP Workshop 2007, France, INRIA, pp. 61–75 (2007)
16. Surynek, P., Barták, R.: Maintaining Arc-consistency over Mutex Relations in Planning Graphs during Search. In: Proceedings of the 20th International FLAIRS Conference, Key West, FL, USA, pp. 134–139. AAAI Press, Menlo Park (2007)
17. Surynek, P.: Solving Difficult SAT Instances Using Greedy Clique Decomposition. In: Miguel, I., Ruml, W. (eds.) SARA 2007. LNCS (LNAI), vol. 4612, pp. 359–374. Springer, Heidelberg (2007)
18. Surynek, P., Chrpa, L., Vyskočil, J.: Solving Difficult Problems by Viewing Them as Structured Dense Graphs. In: Proceedings of the 3rd IICAI Conference, Pune, India (2007)
19. Urquhart, A.: Hard Examples for Resolution. Journal of the ACM 34, 209–219 (1987)
20. Zimmerman, T., Kambhampati, S.: Using Memory to Transform Search on the Planning Graph. JAIR 23, 533–585 (2005)

Author Index

Lecture Notes in Artificial Intelligence (LNAI)

Vol. 5194: F. Železný, N. Lavrač (Eds.), Inductive Logic Programming. X, 349 pages. 2008.

Vol. 5190: A. Teixeira, V.L.S. de Lima, L.C. de Oliveira, P. Quaresma (Eds.), Computational Processing of the Portuguese Language. XIV, 278 pages. 2008.

Vol. 5180: M. Klusch, M. Pěchouček, A. Polleres (Eds.), Cooperative Information Agents XII. IX, 321 pages. 2008.

Vol. 5179: I. Lovrek, R.J. Howlett, L.C. Jain (Eds.), Knowledge-Based Intelligent Information and Engineering Systems, Part III. XXXVI, 817 pages. 2008.

Vol. 5178: I. Lovrek, R.J. Howlett, L.C. Jain (Eds.), Knowledge-Based Intelligent Information and Engineering Systems, Part II. XXXVIII, 1043 pages. 2008.

Vol. 5177: I. Lovrek, R.J. Howlett, L.C. Jain (Eds.), Knowledge-Based Intelligent Information and Engineering Systems, Part I. LVI, 781 pages. 2008.

Vol. 5144: S. Autexier, J. Campbell, J. Rubio, V. Sorge, M. Suzuki, F. Wiedijk (Eds.), Intelligent Computer Mathematics. XIV, 600 pages. 2008.

Vol. 5139: C. Tang, C.X. Ling, X. Zhou, N.J. Cercone, X. Li (Eds.), Advanced Data Mining and Applications. XVII, 759 pages. 2008.

Vol. 5138: J. Darzentas, G.A. Vouros, S. Vosinakis, A. Arnellos (Eds.), Artificial Intelligence: Theories, Models and Applications. XIV, 444 pages. 2008.

Vol. 5129: F. Fages, F. Rossi, S. Soliman (Eds.), Recent Advances in Constraints. VII, 159 pages. 2008.

Vol. 5118: M. Dastani, A. El Fallah Seghrouchni, J. Leite, P. Torroni (Eds.), Languages, Methodologies and Development Tools for Multi-Agent Systems. X, 279 pages. 2008.

Vol. 5113: P. Eklund, O. Haemmerlé (Eds.), Conceptual Structures: Knowledge Visualization and Reasoning. X, 311 pages. 2008.

Vol. 5110: W. Hodges, R. de Queiroz (Eds.), Logic, Language, Information and Computation. VIII, 313 pages. 2008.

Vol. 5108: P. Perner, O. Salvetti (Eds.), Advances in Mass Data Analysis of Images and Signals in Medicine, Biotechnology, Chemistry and Food Industry. X, 173 pages. 2008.

Vol. 5097: L. Rutkowski, R. Tadeusiewicz, L.A. Zadeh, J.M. Zurada (Eds.), Artificial Intelligence and Soft Computing – ICAISC 2008. XVI, 1269 pages. 2008.

Vol. 5081: D. Kapur (Ed.), Computer Mathematics. XI, 359 pages. 2008.

Vol. 5078: E. André, L. Dybkjær, W. Minker, H. Neumann, R. Pieraccini, M. Weber (Eds.), Perception in Multimodal Dialogue Systems. X, 311 pages. 2008.

Vol. 5077: P. Perner (Ed.), Advances in Data Mining. XI, 428 pages. 2008.

Vol. 5076: R. van der Meyden, L. van der Torre (Eds.), Deontic Logic in Computer Science. X, 279 pages. 2008.

Vol. 5064: L. Prevost, S. Marinai, F. Schwenker (Eds.), Artificial Neural Networks in Pattern Recognition. IX, 318 pages. 2008.

Vol. 5056: F. Sadri, K. Satoh (Eds.), Computational Logic in Multi-Agent Systems. X, 299 pages. 2008.

Vol. 5049: D. Weyns, S.A. Brueckner, Y. Demazeau (Eds.), Engineering Environment-Mediated Multi-Agent Systems. X, 297 pages. 2008.

Vol. 5043: N. Jamali, P. Scerri, T. Sugawara (Eds.), Massively Multi-Agent Technology. XII, 191 pages. 2008.

Vol. 5042: A. Esposito, N.G. Bourbakis, N. Avouris, I. Hatzilygeroudis (Eds.), Verbal and Nonverbal Features of Human-Human and Human-Machine Interaction. XIV, 281 pages. 2008.

Vol. 5040: M. Asada, J.C.T. Hallam, J.-A. Meyer, J. Tani (Eds.), From Animals to Animats 10. XIII, 530 pages. 2008.

Vol. 5032: S. Bergler (Ed.), Advances in Artificial Intelligence. XI, 382 pages. 2008.

Vol. 5027: N.T. Nguyen, L. Borzemski, A. Grzech, M. Ali (Eds.), New Frontiers in Applied Artificial Intelligence. XVIII, 879 pages. 2008.

Vol. 5012: T. Washio, E. Suzuki, K.M. Ting, A. Inokuchi (Eds.), Advances in Knowledge Discovery and Data Mining. XXIV, 1102 pages. 2008.

Vol. 5009: G. Wang, T. Li, J.W. Grzymala-Busse, D. Miao, A. Skowron, Y. Yao (Eds.), Rough Sets and Knowledge Technology. XVIII, 765 pages. 2008.

Vol. 5003: L. Antunes, M. Paolucci, E. Norling (Eds.), Multi-Agent-Based Simulation VIII. IX, 141 pages. 2008.

Vol. 5001: U. Visser, F. Ribeiro, T. Ohashi, F. Dellaert (Eds.), RoboCup 2007: Robot Soccer World Cup XI. XIV, 566 pages. 2008.

Vol. 4999: L. Maicher, L.M. Garshol (Eds.), Scaling Topic Maps. XI, 253 pages. 2008.

Vol. 4998: J. Bacardit, E. Bernadó-Mansilla, M.V. Butz, T. Kovacs, X. Llorà, K. Takadama (Eds.), Learning Classifier Systems. X, 307 pages. 2008.

Vol. 4995: A. Artikis, G.M.P. O'Hare, K. Stathis, G.A. Vouros (Eds.), Engineering Societies in the Agents World VIII. XII, 351 pages. 2008.

Vol. 4994: A. An, S. Matwin, Z.W. Raś, D. Ślęzak (Eds.), Foundations of Intelligent Systems. XVII, 653 pages. 2008.

Vol. 4953: N.T. Nguyen, G.S. Jo, R.J. Howlett, L.C. Jain (Eds.), Agent and Multi-Agent Systems: Technologies and Applications. XX, 909 pages. 2008.

Vol. 4946: I. Rahwan, S. Parsons, C. Reed (Eds.), Argumentation in Multi-Agent Systems. X, 235 pages. 2008.

Vol. 4944: Z.W. Raś, S. Tsumoto, D.A. Zighed (Eds.), Mining Complex Data. X, 265 pages. 2008.

Vol. 4938: T. Tokunaga, A. Ortega (Eds.), Large-Scale Knowledge Resources. IX, 367 pages. 2008.

Vol. 4933: R. Medina, S. Obiedkov (Eds.), Formal Concept Analysis. XII, 325 pages. 2008.

Vol. 4930: I. Wachsmuth, G. Knoblich (Eds.), Modeling Communication with Robots and Virtual Humans. X, 337 pages. 2008.

Vol. 4929: M. Helmert, Understanding Planning Tasks. XIV, 270 pages. 2008.

Vol. 4924: D. Riaño (Ed.), Knowledge Management for Health Care Procedures. X, 161 pages. 2008.